# The HANDBOOK of LIGHTING Surveys & Audits

John L. Fetters

CRC Press
Boca Raton Boston London New York Washington, D.C.

Acquiring Editor: Ron Powers
Cover design: Denise Craig
PrePress: Kevin Luong

**Library of Congress Cataloging-in-Publication Data**

Fetters, John L.
The handbook of lighting surveys and audits / John L. Fetters.
p. cm.
Includes index.
ISBN 0-8493-9972-6
1. Commercial buildings--Lighting. 2. Energy auditing.
I. Title.
TK4399.F2F48 1997
621.32′25--dc21 97-36667
CIP


International Standard Book Number 0-8493-9972-6
Library of Congress Card Number 97-36667
Printed in the United States of America 1 2 3 4 5 6 7 8 9 0
Printed on acid-free paper

# Dedication

**This handbook is dedicated to my father, Donald Fetters, and to my father-in-law, Elton Robertson, who would have been pleased that such practical information has been provided for use by workers in the field.**

# Preface

Companies are actively upgrading their lighting systems. The forces driving this activity include programs such as EPA Green Lights, utility Demand Side Management, and government programs to reduce energy in federal facitlities. These programs target lighting to reduce electrical energy. Corporate office and manufacturing spaces are being upgraded to improve worker productivity and company profitability. Older retail environments are being renovated to attract new customers. All this upgrade activity requires a complete lighting evaluation of existing systems.

The first step of any lighting evaluation requires performing a lighting audit. The second step—identifying opportunities for savings and recommending equipment, and the third step — calculating savings and potential payback — are substantially affected by the information collected in the lighting audit. A casual approach to the audit process can result in costly mistakes later. To avoid these mistakes requires an investment in time and effort spent on performing an effective lighting audit.

The objective of this practical handbook is to provide the auditor with solid, practical information to help perform building lighting surveys and audits. With the advice given here, the reader will learn how to conduct surveys and audits more accurately and more efficiently. This handbook can be used as a text for the beginning surveyor as well as a valuable field source for the experienced lighting auditor. The purpose of this handbook is to give all readers a better understanding of the complete survey and audit process, as well as practical tips and techniques that will speed the process.

Several checklist forms are provided that can be copied and used for building lighting surveys and audits. Sample reports are shown in the format that they were provided to clients to illustrate many of the points made in the book. Audit tools and measuring instruments are described in detail, and the auditor is guided in their proper use. Lists of manufacturers, their address, and phone numbers are provided.

Lighting Energy Conservation Measures (ECMs) are listed, along with descriptions and caveats for their recommended use in upgrading building lighting systems. Concise Fact Sheets guide the reader through a myriad of important lighting details and an extensive glossary defines all important lighting terms. Abbreviations and acronyms used in the book are expanded for better understanding. Commercial software, available for audits and analysis, is described, and their use is explained in detail.

# The Author

**John L. Fetters, CEM, CLEP,** is an accomplished energy engineer with special expertise in energy efficient lighting. He received a BEE from Cleveland State University and the Energy Management Diploma from the University of Wisconsin. He has been a Certified Energy Manager since 1984 and was awarded Certified Lighting Efficiency Professional status in 1992. He is a qualified EPA Green Lights Surveyor Ally and in 1994 passed the IESNA Technical Knowledge Examination (TKE).

John is a member of the Illuminating Engineering Society (IESNA), president of the IES Buckeye Section, and serves on the local Education Committee. He has served as a judge for the IESNA International Illumination Design Awards program, is an active member of the IESNA Industrial Lighting Committee, and is currently serving on the LIGHTFAIR '98 program committee. John has received three International Illumination Design Awards from the IESNA.

John is a Senior member of the Association of Energy Engineers (AEE), is regional vice president, and chairs the Certified Lighting Efficiency Professional board. He was distinguished with the AEE Energy Professional Development award in 1993.

John has published more than 65 technical articles and books and continues to contribute his expertise in lighting on the Energy User News (EUN) Technical Advisory Board. He is editor of the widely acclaimed EUN energy management training series and received first place in the 1996 Chilton Editorial Award for the best how-to article. Most recently he has written articles for *Maintenance Solutions* magazine and the update for Chapter 9 (HID Lighting) for the *E SOURCE Lighting Atlas*. He chairs the IESNA subcommittee to edit Chapter 20 (Industrial Lighting) of the *IESNA Handbook*.

John is an accomplished speaker presenting seminars and technical sessions at major lighting, energy management, electric, and power quality conferences. He has developed and taught all the Aee lighting seminars since 1989 and lighting courses offered by the Buckeye Section of the Illuminating Engineering Society (IES).

He is a regular guest lecturer for the University of Wisconsin, teaching "Lighting Systems and Controls," "Lighting Evaluations and Audits," "Computer Calculation Tools," "Fundamentals of Energy Audits," and "Building Energy Systems."

# Contents

# Chapter 1
# Introduction to Lighting Surveys and Audits

## Background

Lighting is a principal user of electrical energy in commercial buildings, accounting for about 30% of commercial sector consumption. For buildings using newer lighting technology, lighting consumption can be reduced realistically by approximately 30 to 50%. This represents a significant opportunity to save energy in commercial buildings and to reduce the environmental emissions that accompany the electrical generation process. Virtually all facilities can be upgraded with advanced hardware to increase efficiency and improve lighting quality.

More important, there is a lot of old, ineffective lighting in schools, offices, manufacturing, and even some retail spaces. Although many use the terms efficient and effective interchangeably, there is a difference, and for someone whose task is to improve lighting systems, understanding the distinction is important. Efficient lighting systems can be described in engineering terms of performance and measured in lumens per watt and watts per square foot. Effective lighting is described in terms of how appropriate the lighting is for the function it is performing. In this usage, effective means doing the right thing. An analogy may help. Someone can work very efficiently, being very productive in a job, but if that person is not doing the right things or setting the right priorities, then the job is not being done effectively. This concept is sometimes summarized as "Work smarter, not harder."

Lighting systems need to be evaluated not just in terms of lumens and watts, but also by means of a broader approach that encompasses the visual tasks and the environment to be lighted. The premise of this handbook is that lighting professionals are responsible for doing both types of evaluation and then making recommendations that will result in both efficient and effective lighting systems. The need for balance between efficiency and effectiveness is emphasized throughout.

## Lighting Efficiency Programs

### Demand-Side Management Programs

Since the 1980s, lighting efficiency programs have been popular with electric utilities to improve overall efficiency of electricity use and as a low-cost alternative to building expensive power plants. These demand-side management (DSM) programs usually paid rebates directly to the customer. The most common method was to pay a fixed amount for each installed lighting component (lamp, ballast, EXIT light, etc.) called rebates. Utility companies usually calculated customer payments based on estimated demand since they did not have verifiable data to document actual DSM impacts. Later, in the 1990s, utilities used customer polls, metering programs, and computer modeling to more accurately approximate DSM impacts. As a result, programs became more sophisticated and DSM became an accepted piece of utility integrated resource planning (IRP). IRP considers a wide range of both supply-side and demand-side options to meet future customer electrical energy needs. Approximately 30 states required utilities to consider DSM as part of their resource mix, and many state public utility commissions (PUCs) employed measures that allowed utilities to earn a return on their DSM program investments.

Spending on DSM programs increased dramatically between 1989 and 1993, increasing from 0.5 to 1.5% of utility operating revenues. DSM programs were responsible for reducing electricity use and peak demand by 44,000 GWh and 40,000 MW, equivalent to 1.6% of annual electricity sales and 6.8% of summer peak demand. In 1994, DSM spending began to slow; however, lighting maintained its position as a key DSM approach. In 1994, utilities sponsored 1969 DSM programs, and lighting represented 332 (16.8%) of these. Some analysts recently have questioned the cost-effectiveness of lighting DSM programs, but many lighting programs continue to prove cost-effective.

Successful lighting DSM programs have helped transform lighting markets. Electronic ballasts have captured a 25% market share in the less than 15 years since their introduction to the market. Utility DSM programs that aggressively promoted this technology in the commercial sector contributed to the technology's widespread availability. DSM programs have also increased customer awareness of the benefits of lighting. Although customers initially were skeptical, many now indicate that they became proponents of lighting energy efficiency as a result of their participation in a DSM program and also indicate that they would be willing to install such measures without utility rebates. Participating customers also talked with nonparticipants, indirectly influencing their attitudes about the new technologies and extending the direct benefits to an originally nonparticipating constituency.

Recent trends show a shift in how utilities are structuring their DSM programs. In general, utility programs will change to become more cost-effective and service-oriented. Utilities that have aggressively implemented DSM are likely to continue in this mode in order to use their least-cost mix of resources and to defer transmission and distribution (T&D) expenditures. Many utilities will use DSM to retain customers by offering programs that are valuable enough to help keep existing customers on their system. These utilities will shift their emphasis away from residential customers, focusing more on commercial and industrial (C&I) customers. Some programs will emphasize value-added services such as equipment leasing, and new economic incentives such as financing, shared savings programs, and creative energy pricing. Rebates will continue to decrease in popularity.

For lighting auditors, the utility sector activity in lighting surveys and audits is expected to at least hold steady, and it could increase. Most of the utility surveys will require auditors who work for independent lighting companies as utility staffs continue to be downsized. The utility market is shifting and the perceptive auditor will track these shifts to take advantage of them.

### EPA Green Lights Program

The EPA Green Lights program is a voluntary, nonregulatory program aimed at promoting energy efficiency through investment in energy-efficient lighting. The program saves money for businesses and organizations and creates a cleaner environment by reducing pollutants released into the atmosphere. Green Lights is designed to reduce pollution, to promote public-private partnership, to use market forces, and to recognize environmental leadership by encouraging participants to install energy-efficient lighting.

Participants benefit with lower electric bills, improved lighting quality, an enhanced corporate image, and a cleaner environment. Participating companies are asked to sign a Memorandum of Understanding (MOU), survey all domestic facilities, upgrade lighting wherever profitable, and complete upgrades within five years. EPA Green Lights supports participants with state-of-the-art software to support decision making (see Chapter 6), a financing database, "consumer reports" of lighting products (via The National Lighting Product Information Program, NLPIP), and networking with lighting manufacturers, management companies, and utilities.

More than 2200 participants have committed 5.6 billion square feet of their buildings as of July 1996. Of that amount, more than 1 billion square feet have been upgraded to energy-efficient lighting so far. Participating companies, organizations, and government agencies have invested $930 million, resulting in nearly 3 billion kWh electrical savings per year ($225 million per year). The average electrical reduction for completed upgrades is 48%.

## Energy Policy Act of 1992 (EPAct)

The first comprehensive energy policy for the United States was signed into law by President George Bush on Oct. 24, 1992. Major provisions of EPAct establish important energy efficiency standards, encourage greater competition in electrical generation, and provide tax incentives for domestic energy production and conservation.

In addition to the establishment of new statutory standards for appliances and certain equipment, such as electric motors and lighting, EPAct encourages electric and gas utilities to consider IRP and DSM programs. The act gave the federal government some additional tools for energy management, including training and audits.

## Federal Energy Management Program (FEMP)

FEMP came into being in the mid-1970s in response to the oil embargo, as many in the federal government believed it should try to manage its own energy use better. Through FEMP, the Federal Department of Energy (DOE) worked to provide leadership; to improve efficiency and fuel flexibility, not only in buildings but also in transportation and industrial process energy; and to facilitate the transfer of the developed knowledge to other federal agencies. There are about 500,000 buildings in the federal sector at 29 agencies that report to FEMP. There are approximately 8000 federal sites where the government owns and operates buildings and thousands more where it leases space. The federal government uses more energy than any other individual group in the country, and federal buildings use about 40% of the government energy budget. Taxpayers paid approximately $11.2 billion for that energy in 1992. FEMP energy savings goals have been realized, achieving a 10% reduction from 1985 to 1995. Another 10% reduction is forecast for the period from 1995 to 2000. EPAct affirmed the 20% savings by the year 2000 that President Bush had ordered.

Presidential Order 12902, signed by President Clinton in 1994, increased the energy reduction goal to 30% from the base year 1985 and extended the completion date to the year 2005. Industrial efficiency improvements are included to reduce energy by 20% between 1990 and 2005. Other provisions of

Presidential Order 12902 require that prequalified federal buildings be audited at the rate of 10% per year. A large part of these energy audits will include evaluating lighting systems.

## The Changing Role of Utilities

Electrical utilities are positioning for an unregulated, more competitive business environment, partly brought about by EPAct and partly because of a deregulation trend. Many U.S. business sectors, such as telecommunications and airlines, have already been deregulated. Utilities will change in size and will play different roles as deregulation becomes a reality. Evidence of this trend is seen daily in news accounts of utility mergers. The transition from regulation to competition has larger companies in pursuit of weaker ones. Mergers have been rare in the electric utility industry until recently, but from January 1994 to July 1996, 14 deals worth $47 billion and involving 30 companies were announced. By the year 2000, it is estimated that mergers will have reduced the number of publicly traded utilities from more than 100 to fewer than 80. Mergers usually increase efficiency and that benefits both investors and consumers. Well-run companies have stronger earnings and can offer customers the best deal. The price for this is a loss of flexibility when the utility becomes larger. Value-added services will be offered to existing customers to keep them on the utility's lines. Some of these services may involve lighting, providing another market for the lighting auditor.

## Opportunities for Survey Work

Of particular interest to lighting auditors is the Green Lights requirement that all Green Lights Partners survey each of their domestic facilities. With a total of 70 billion square feet of office space alone in the United States, the work has just begun. In addition, retail stores, industrial facilities, government spaces, hospitals, and schools will all need to be surveyed in the next several years.

Firms which perform lighting audits and surveys which are not presently Green Lights Surveyor Allies should plan to become Surveyor Allies as part of their strategic business planning. Candidates attend a two-day Green Lights Surveyor Ally preparation seminar and to qualify must pass a test given at the end of the seminar. After passing the test and signing a MOU, a candidate is then recognized as a Surveyor Ally. Auditors who promote Green Lights can improve their business by developing new surveying opportunities.

Although more lighting surveys have been accomplished in the past few years than ever before, according to the U.S. Department of Energy's Energy Information Administration (EIA), only 22% of commercial floor space has had an energy audit since 1987. This means that there will be even more surveys and audits needed in the future as more and more facilities are included.

The scope of potential survey work can be seen in another way. Using figures from the fluorescent ballast industry, the potential can be measured in hundreds of millions of luminaires. There are 1.6 billion ballasts in the United States as of mid-1996, but only about 200 million have been replaced in conjunction with utility or government incentive programs. This is good news for auditors, who have several years of work ahead of them.

There also is a need for lighting audits in a large percentage of those 500,000 buildings in the federal sector. Auditors who have demonstrated qualifications through certification such as the Certified Lighting Efficiency Professional (CLEP) from the Association of Energy Engineers or the Certified Lighting Management Consultant (CLMC) from the National Association of Lighting Management Companies (NALMCO) should be highly regarded when this work is contracted.

## Lighting Upgrade Approaches

To reduce energy costs, building owners have started to retrofit their building lighting systems. In many cases, a quick-fix, single-technology component approach has been used that has fallen short of the full benefits possible with a more comprehensive lighting management approach.

One deficiency of the upgrade practices employed to date can be seen in the improper light levels being used, even in retrofit systems. Scores of systems are not providing the proper light levels for the tasks they light, and some are providing higher than recommended levels. The reason for these mismatches is that when these lighting systems were installed they were put in place to light tasks that have long since been changed and, even if these systems were retrofit, the lighting may not have been changed to accommodate the different tasks.

Another deficiency is the perceived lack of quality in many lighting retrofits structured on energy savings alone. In particular, early specular reflector projects resulted in a perceived poor quality by occupants. The problem was not that reflectors were used, but that they were not used correctly and that those using them did not understand luminaire photometrics. The attractiveness of low first cost drove the selection of commodity-grade reflectors. These "one size fits all" inserts could be put in nearly all four-lamp fluorescent 2 x 4 (2' x 4') troffers, and two lamps were removed. In most cases, the remaining two lamps were not relocated and the spacing criteria (SC) of the unit was severely reduced. When the SC of luminaires already mounted on wide centers is reduced, the effects are noticeable on walls and are less noticeable between luminaires. The point at which the light intercepts the wall is lowered when the SC of the units closest to the wall is reduced. The result is that the walls appear darker. The solution is either to use custom reflectors or provide an asymmetrical reflector in those luminaires located in rows closest to the walls. Most custom reflector designs do not reduce the spacing criteria. The point is, using commodity reflectors resulted in dark walls and a dimmer environment, and most occupants did not believe that anyone had improved their work space.

A similar problem surfaced with one-for-one retrofits of 4-foot T8 lamps for 4-foot T12 lamps. The T8 lamp, being smaller in diameter and the same lumen output as the T12, has a higher surface brightness (luminance). In addition, because the color rendering index (CRI) of the T8 lamp is much higher than the CRI of the T12, the T8 appears brighter. Many projects suffered a high number of occupant complaints after this type of conversion, but in those projects in which the number of lamps was reduced from four to three, occupants were generally satisfied with the lower light level.

These are just two examples of how lighting quality can be compromised when the concentration is on energy savings and quality is left to chance. Quality lighting needs to be designed into upgrade or relighting projects for the most effective results.

The broader, lighting management approach not only takes into consideration proper light levels and lighting quality, but also starts with the visual needs of the people who work in the spaces. To the enlightened, the lighting audit is a tool for improvement. In the language of total quality management, continuous improvement — what Japanese management calls kaizen — should be both the reason for conducting a lighting audit and the primary goal that is set as a result. No equipment specifications should be written or analysis performed without complete information about the visual tasks being performed, the existing spaces, and the lighting systems and their operation. This information is absolutely necessary before changes can be proposed to improve the working environment.

## Two Views of Lighting Systems

Many building owners and energy managers view lighting systems as just one of several large building systems that use too much energy, produce too little light and suck up a large portion of the maintenance budget. In this view, the lighting is seen as a large part of the overhead cost of operating the building.

More enlightened owners and managers view the lighting systems as a key component of a quality workplace and realize that quality lighting plays a vital role in supporting worker performance and satisfaction and helps project a positive company image. In quality manufacturing environments, lighting is seen as an important input to the production process.

In the limited view — lighting as part of the overhead cost of operations — lighting is seen as bright hardware that negatively impacts the bottom line, wasting important corporate resources and incurring high costs that need to be controlled. The enlightened view sees lighting systems as significant resources that play a pivotal role in creating productive corporate environments and enhancing the company image.

When the focus is only on a building's lighting systems, the people who work in the building — who are, after all, the reason for the lighting — become secondary or fade out of view altogether. Total concentration on the lighting hardware makes it easy to miss opportunities to provide needed improvements in the lighted environment. Although productivity gains and employee satisfaction are difficult to measure, they far outweigh energy and maintenance savings.

The pie chart in Figure 1.1 shows the average office costs per square foot in the United States. The energy cost of $3.34 per square foot represents approximately 2% of total office costs. Lighting is part of this piece of the pie and is estimated to be less than 1% of total office costs, making it the lowest of all the office costs.

When people become the focus of lighting improvements and when modern lighting technologies are brought to bear, the results will include not only energy and maintenance savings but also an improved working environment that will contribute to increased worker productivity — a comprehensive solution.

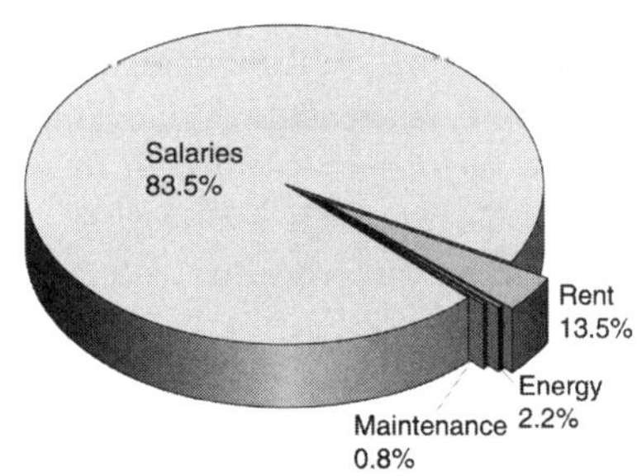


FIGURE 1.1 Average U.S. office costs per square foot. Source: Columbia Lighting.

## The Lighting Survey Defined

Dictionaries tend to give two definitions of the word survey (used as a noun): "a detailed inspection" and "a general or Comprehensive view." Leaning toward the second sense, in this handbook we use the term survey to describe an examination process that is simpler than an audit and that is characterized by a walk-thru with a minimum of measurement or analysis.

## The Lighting Audit Defined

The *American Heritage Dictionary* defines the verb form of the word audit thus: "To examine, verify, or correct the financial accounts." Not surprisingly, this sense of the word has disagreeable connotations for some, conjuring up the specter of government audits and, in particular, the legendary audits of the Internal Revenue Service. Because of these negative associations, some lighting professionals will not even use the word audit to refer to lighting evaluations and instead choose to use the term survey.

*Compton's Interactive Encyclopedia* emphasizes that an audit usually is conducted by "specialists other than those responsible for their preparation." While the term audit was first associated with financial conditions, it can also describe any "thorough investigation." In this handbook we use the term lighting audit to describe a detailed, systematic evaluation of the existing conditions of lighted spaces and the performance of lighting systems. The audit is characterized by detailed data collection, measurements, and in-depth analysis of the data, usually performed by third-party specialists.

## Purposes of Surveys and Audits

The main purpose of a lighting survey or audit is to gather information concerning the characteristics and the current condition of lighting systems and the lighted environment. This information is used as baseline data. The data collected becomes a "snapshot" in time that may be compared with (1) standards of performance, (2) a proposed upgrade or relighting plan, or (3) later performance.

Another purpose may be to help provide the justification for upgrading the lighting systems. This information is usually used to quantify the potential monetary savings that may be realized from upgrading but also be may used to justify other potential benefits such as employee productivity.

A third purpose of a lighting survey or audit may be to obtain information for a project feasibility study. This information is used to determine if a relighting or upgrade is possible within the constraints imposed by the building owner or operator. This information usually is provided at the concept stage of a proposed project and may include data that will help determine the scope of the project. Ideally, this data should be convincing evidence for decision makers that the project scope should include a lighting upgrade because, in addition to all the other benefits, including the lighting upgrade will help pay for the rest of the project. ESCOs (energy services companies) and performance contractors often will use the lighting upgrade as the base for their proposed project to carry their other building energy improvements.

In some cases, lighting surveys and audits are used to break down projects in a priority order. The information gathered is used to prioritize lighting projects or larger energy management or remodeling projects. The priority order criteria are supplied by the building owner or operator and the projects are usually broken down by building area or by payback period.

A lighting survey or audit should be part of any plan to modernize work spaces. As part of the modernization process, the audit supplies data to determine costs and benefits of upgrading or relighting when renewing finishes and installing new carpet or office furniture. In the case of manufacturing facilities, it may be done when an area gets a new or improved process or assembly line. Quality circle groups can supply valuable information concerning the important role that lighting plays in their jobs and even how they think that the lighting may be improved. The lighting professional should know how to employ task lighting for these specialized situations.

Occasionally, surveys or audits are performed after an upgrade is installed. This survey or audit is performed either where commissioning is required by a building owner or property manager, performance contractors, or sponsoring electric utilities. The reason for performing the survey or audit after a relighting or upgrade is to verify performance and compare it with the baseline condition.

## Types of Surveys and Audits

As is discussed in detail in Chapter 3, there are three types of surveys and audits: the walk-thru survey, the intermediate audit (mini-audit), and the comprehensive audit. These three types form a hierarchy from the walk-thru survey (simple) to the comprehensive audit (complex). The intermediate survey is used for those situations too complex for a walk-thru survey and too simple for a comprehensive audit. Note that the Green Lights survey mentioned earlier in this chapter may be considered an intermediate audit.

The purpose, reason, or need for performing a survey or audit will determine which type is required. If the purpose is to establish if there is potential for a lighting upgrade, or if the owner simply wants a "checkup" of the lighting systems, then a walk-thru survey usually will accomplish these purposes. On the other hand, a comprehensive audit will be used if a facility needs an extensive relighting. Also, the more extensive the analysis required, the more data will be needed, and thus the more comprehensive the audit conducted. Cost may dictate the type of survey or audit. Surveys are less time-consuming and therefore less expensive than the more time-consuming comprehensive audits. However, the investment in a comprehensive audit can pay dividends because the amount of data collected and the number and scope of improvements recommended are proportional to the effort and cost of the survey or audit.

### The Walk-Thru Survey

The walk-thru survey is the simplest type and is usually performed while the auditor simply walks through a facility (hence the name). A minimum amount of data is usually collected by one- or two-person teams. There is little analysis beyond simple payback, and a brief report describes the existing systems, outlines the proposed improvements, and reports the estimated payback.

### The Intermediate Audit (Mini-Audit)

The intermediate audit is a hybrid of a walk-thru survey and comprehensive audit, containing elements of each. It is more detailed than a walk-thru survey, and the data collection is usually done by a small team of two or three auditors.

The financial analysis made for the intermediate audit usually is more than simple payback, but it still may not include a full-scale life-cycle costing procedure. Calculations are usually performed for the complete project instead of on a fixture or component basis. Reports usually include equipment inventories, power densities, limited evaluation of upgrade or relighting alternatives, and payback.

### The Comprehensive Audit

The comprehensive audit is performed when the highest level of detail is required and is usually conducted by small teams of two or three. At this level the auditors focus on visual tasks and the worker's productivity as well as on lighting equipment performance. All lighting equipment is fully identified and counted, and a detailed evaluation of lighting quality improvements is outlined. The assessment of daylighting potential is optional, depending on the purpose and scope of the audit. An extensive analysis is performed, including life-cycle costing.

A comprehensive report describes the condition of existing lighting systems and outlines several upgrade options, including relighting, with an analysis of each. Both positive and negative aspects of each measure considered should be reported so that the client can make an informed choice. Maintenance savings are calculated and reported, as well as the potential for productivity improvements. A description of the new look of the upgraded or relighted spaces may be included, supported by drawings or perhaps even rendered using a computer program with an advanced graphics package.

## Summary

Efficient lighting systems can be described in engineering terms of performance and can be measured in lumens per watt and watts per square foot. Effective lighting is described in terms of how well or how appropriately the lighting is matched to the function it is performing. In this usage, effective means doing the right thing.

The main purpose of a lighting survey or audit is to gather baseline information concerning the characteristics and the current condition of lighting systems and the lighted environment. The data collected become a "snapshot" in time that may be used to compare with (1) standards of performance, (2) a proposed upgrade or relighting plan, or (3) later performance. Other purposes for performing a lighting survey or audit are (1) to help provide the justification for upgrading the lighting systems by quantifying the potential savings that may be realized from upgrading, (2) to obtain information for a project feasibility study by determining if relighting or upgrade is possible, (3) to break down projects in priority order by using lighting audit analysis to prioritize projects, (4) to supply data to determine costs and benefits of upgrading or relighting when remodeling an office or when a manufacturing process or assembly line is added or changed, and (5) after an upgrade is installed, to verify performance and compare with the baseline condition.

Demand-side management (DSM) programs are energy efficiency programs put in place by electric utilities to improve overall efficiency of electricity use and as a low-cost alternative to building expensive power plants. DSM programs have been successful since their introduction in the 1980s, but they may be in decline due to the restructuring of the electric power industry.

The provisions of the Energy Policy Act of 1992 (EPAct) and related presidential orders have had an impact on lighting surveys and audits by encouraging electrical efficiency, competition in electrical generation, tax incentives, and the like. The Federal Energy Management Program (FEMP) has had an impact, too, as federal buildings must strive to meet stated energy-reduction goals. These programs, along with the Green Lights programs of the EPA, point to increasing opportunities for lighting auditors.

The quick-fix, single-technology approach to lighting upgrades falls short of the full benefits possible using a more comprehensive lighting management approach. Similarly, the view of lighting systems as an element of overhead cost that negatively affects the bottom line falls short of the benefits produced by an enlightened view that sees the lighting systems as key components of a quality workplace, playing a vital role in supporting worker productivity and satisfaction.

A lighting audit is a detailed, systematic evaluation of the existing conditions of lighted spaces and the performance of lighting systems. The audit is characterized by detailed data collection, measurements, and an in-depth analysis of the data, usually performed by third-party specialists. A survey, on the other hand, is a relatively simple examination characterized by a walk-thru approach with a minimum of measurement and analysis. The three basic types of surveys and audits are the walk-thru survey, the intermediate audit (mini-audit), and the comprehensive audit. These three types form a hierarchy from the simple walk-thru survey to the comprehensive audit, which is the most complex. The intermediate survey is used for those situations that are too complex for a walk-thru survey and too simple for a comprehensive audit.

# Chapter 2
# The Lighting Evaluation Process

## Evaluating Lighting Systems

To evaluate lighting systems means to examine or carefully appraise them. Evaluating lighting systems is a multistep process. At the very minimum, the evaluation is a three-step procedure. Step 1 of any lighting evaluation requires performing a lighting survey or audit to establish a baseline of performance. Note that the audit could also be the last step, as discussed in Chapter 1.

Identifying opportunities for improvements (Step 2) and calculating savings and potential payback (Step 3) are substantially affected by the quality of the information collected in the lighting audit. A casual approach to the survey or audit step can result in costly mistakes later. To avoid these mistakes requires an investment in time and effort spent on performing an effective, accurate lighting survey or audit.

## Step 1: Performing a Lighting Survey or Audit

The characteristics of each lighting system need to be assessed, including operating conditions, operating hours, and maintenance. This assessment is provided by a lighting survey or audit. As explained in Chapter 1, the purpose, reason, or need for performing a survey or audit will determine which type of survey or audit is required for each project.

Once the facts are determined for each lighted space, objectives can be set for improvements and an action plan can be formulated. The action plan resulting from a walk-thru survey is a simple report that outlines the recommendations. For a comprehensive audit, the action plan may be in the form of a lighting management plan. Lighting management is much broader than lighting energy conservation since it also includes the quality of the lighted environment, visual performance, and company priorities. The lighting management plan is a plan of actions that guides the lighting upgrade or relighting process.

## Step 2: Identifying Opportunities for Improvements

Opportunities are advantageous circumstances or combinations of circumstances uncovered by the auditor. Training and, eventually, experience will teach the beginning auditor how to search for and discover those circumstances that can be turned into opportunities for improvements.

Improvements are the changes that raise the existing condition of the lighted environment to a more desirable condition or to a more excellent quality, that is, that make the lighted environment better. What to change and how to change will depend on the particular focus of the analyst and the objectives of the upgrade or relighting project.

There are several bases for identifying opportunities for improvements. They can be categorized in terms of quantitative and qualitative identification.

### Quantitative Identification

There are three measurable bases for identification of opportunities for improvements: power density (PD), light levels (illuminance), and lighting controls.

#### Power Density (W/sf)

The first basis for identifying opportunities for improvement is power density (PD). Power density is the power allowance for a building, measured in watts per square foot (W/sf). The calculation of power density is detailed in Chapter 8, Fact Sheet 13.

The efficiency of installed lighting systems can be determined by using power density as a figure of merit. For example, in the current lighting practice of the late-1990s, new, efficient office lighting systems are measuring 1.0 W/sf for systems without daylight contribution. Office lighting systems that calculate above this value should be considered inefficient and are good candidates for upgrade or relighting.

A revised draft of the ASHRAE/IESNA 90.1 standard, in public review at midyear 1996 and expected to be approved sometime before 1999, revises the power density values downward to better reflect state-of-the-art lighting technology and also mandates lighting controls. With the Energy Policy Act of 1992 (EPAct; see Chapter 1) strongly recommending that the states codify 90.1, new designs will be required to meet the new power density code.

The power density of lighting systems can be used by the auditor as a screening measure. If a particular lighting upgrade is calculated to be cost-effective for spaces of a certain power density, then all similar spaces measuring higher than that power density can also be upgraded cost-effectively.

### Light Level or Illuminance (FC)

The second basis for identifying opportunities for improvement is light level. Existing light levels are either appropriate for the lighted environment and the current tasks they are lighting or they are not. Measuring illuminance during the audit will allow the analyst to determine if levels are appropriate for each space by comparing the levels with current Illuminating Engineering Society of North America (IESNA) recommendations. Chapters 4 and 5 outline the tools and techniques used to measure illuminance.

How much light is required to illuminate work spaces or to perform tasks? The IESNA illuminance recommendations are the best guides. These consensus recommendations have been agreed upon by lighting professionals. The IESNA *Lighting Handbook*, 8th edition (the 9th edition is in preparation and is scheduled for publication in 1999) contains the current illuminance selection procedure and the latest illuminance recommendations for different types of spaces and for various tasks. However, there are no single value recommendations for any particular space or specific task.

A host of pertinent issues must be given careful consideration before any conclusion is reached regarding the appropriateness of measured light levels. These include, but are not limited to, the type of space and how the space is used, the flexibility of use, the desired lighting effects, the visual tasks, the owner's liability, and several weighting factors such as the age of the occupants, the importance of speed and/or accuracy in performing a given task, and surface reflectances. These judgments are best made by an experienced lighting professional.

Proper use of the IESNA illuminance procedure and recommendations will help avoid overlighting. In spaces the analyst determines are overlighted, lighting requirements must be changed to provide a lower level of light. Overlighting not only wastes energy but also reduces lighting quality. To improve these overlighted spaces, recommendations are made in the report that will provide a calculated lower light level. (See Chapter 8, Fact Sheet 13 for zonal-cavity calculations.)

For spaces that are found to be underlighted (a common occurrence in stairwells, even though owner liability in such places is high), the lighting professional must ensure that sufficient light is recommended for the safety and security of occupants and visitors. In fact, in the upgrade recommendations report, a list of areas identified as underlighted safety situations should be included up front. Reporting underlighted areas and recommending their replacement or additional lighting will probably increase the power density; however, highlighting these safety concerns as the highest-priority recommendations will strengthen professional credibility and provide the proper notification of safety problems to the building owner or manager. Failure to notify of unsafe conditions uncovered during a survey or audit can become a serious legal liability for the lighting professional.

The issue of lighting ceilings and walls is sometimes judged "wasteful" in articles dealing with energy management of lighting systems. Sometimes the advice is to "only put light where it is needed." However, room surfaces need to be appropriately lighted to provide a space with luminance for several reasons: (1) to define the space, (2) to balance the luminance ratios in a space, and (3) to prevent having a cavelike environment. Balanced lighting will provide a quality lighted environment in which people can do productive work.

### Lighting Controls

Lighting controls form the third basis for identifying opportunities for improvement. Lighting systems that are not controlled waste energy and project a negative image of the company or organization.

Most older lighting systems were installed with no visible means of control, often relying on protective circuit breakers for on/off switching. When circuit breakers that are not rated for switching duty are used as switches, they can fail to function as protective devices when an electrical overload situation occurs. This condition should be highlighted as a safety concern.

The absence of effective lighting controls provides the opportunity for improvement. Identifying this need during a survey or audit is straightforward and only requires that the auditor record or enter in a checklist the type of control (including none) for each area. The lighting professional providing the analysis will determine from all the data collected what type of controls to recommend.

#### *Switching Opportunities*

The effect of switching on lamp life is shrouded in mystery and mythology. Two myths, developed in an earlier time of cheap electricity and original fluorescent lamp characteristics, have kept many companies from installing switching to control lighting systems. The first myth states that continuously operating fluorescent lights is cheaper than turning them off for brief periods. The second myth states that lights should not be turned off because this shortens lamp life and increases maintenance costs. These effects are explained in detail in Chapter 8, Fact Sheet 12. With technology advancements and increased electricity costs, the traditional wisdom of operating fluorescent systems for long periods of time can no longer be supported.

The auditor records the switching methods used in each space. This information, along with the description of the tasks performed in a space, will allow the analyst to determine what potential opportunities exist for controlling the lighting systems. Some areas will best be controlled by occupancy sensors, while others will be better served by timers or central control systems. No one particular control solution will be right for all areas. These control decisions need to be made judiciously by an experienced lighting professional.

#### *Dimming Opportunities*

Recently, the option of changing light levels in individual spaces has become available through electronic dimming equipment. Electronic dimming ballasts combined with preset activity-based controls are now affordable and can provide the option of controllable light levels to match the changing space use requirements. For some applications, it may be preferred by the occupants that in unoccupied spaces the lighting be dimmed to a low level instead of being turned off, especially in large open office areas.

Dimming equipment controlled by photosensors can be used for daylighting controls to prevent spaces from being overlighted with combinations of electric lighting and daylight. The auditor will be aware of and record those opportunities for dimming control.

### Qualitative Identification

Qualitative identification of opportunities requires subjective judgments of lighting systems. Experience is the best preparation for making subjective judgments, and even then the reader needs to realize that this is an area where there can be passionate discourse: this is the stuff lighting professionals argue about. Since none of us sees exactly the same way and none of us has the precise same acceptance level, or the same taste, it is no wonder that we argue about what is "good" lighting. In spite of all this, experienced auditors should be able to identify lighting systems that are in need of quality improvement.

Lighting quality is complicated, and there is no single measure like illuminance (a quantitative measurement) to measure the quality of light. Lighting quality considers visual aspects that are highly subjective and not easily quantified. We can, however, understand and evaluate several quality attributes or characteristics, including veiling reflections, glare, color and CRI, lamp flicker, luminance ratios, and visual comfort probability. Each of these quality attributes is discussed in detail in Chapter 8, Fact Sheet 1.

## Step 3: Calculating Savings and Potential Payback

### Lighting Economics

Lighting costs can be broken down into first costs and operating costs. There are three elements of operating costs: energy, labor, and lamp cost. The relative cost of each is shown as a percentage of the total operating cost in Figure 2.1, with electricity at 86%, labor of 11%, and lamp cost at 3%. Figure 2.1 helps explain why it is almost always cost-effective to replace a lamp with a more efficient (higher-efficacy) one, even though the new lamp may be more expensive. With the lamp cost relatively small compared with the electricity cost, there is a payback for improving lighting efficiency.

#### Electricity Costs

Electric utility bills contain many cost components that the analyst needs to know. The most important components of monthly billings include energy usage, in c/kWh: peak demand charges, in $/kW; fuel adjustment, in 4/kW; and power factor penalty (sometimes called excess kilovolt-amps [kVa]). The analyst should be able to calculate the effect of a lighting upgrade on the client's electric bill without using average rates.

#### Electric Rate Structures

When savings are calculated using average savings or "national averages," the true savings will be distorted. Use of average savings can lead to either overestimating or underestimating true savings. These results can mislead a client to either prolong performing an upgrade or proceed with an upgrade with an expected savings that cannot be realized.

Most commercial-industrial rates have a peak demand charge as one of the highest components of the billing, sometimes accounting for more than half of the electric billing. The peak demand charge is for the maximum rate at which energy is used (measured in

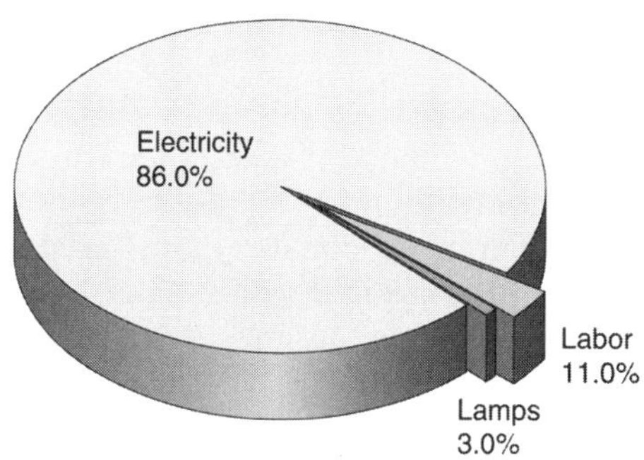


**FIGURE 2.1** Electricity, labor, and lamp costs as a percentage of lighting operating costs. *Source*: Philips Lighting.

kWh/hr or kW). The analyst needs to be sure to include the effect on the demand portion of the billing. When energy-efficient lighting upgrades are installed in interior spaces, the lighting power load is reduced, reducing peak demand. This savings can be larger than the savings calculated using average rates, so it is imperative that demand be included in the calculations. On the other hand, exterior lighting that is operated at night, usually during off-peak hours, will have a smaller effect on the electric bill. Historical demand profiles, which are available from electrical utilities for the audited facility, are useful to the analyst.

Declining block rates are not as prevalent as they once were, but they can be the source of large errors if they are not understood. In declining block rates, the more electricity used, the cheaper the rate. Savings calculated on this rate must be calculated at the lowest block rate, because *the first kWh saved is the last kWh purchased.* Using average rates in this case always overestimates savings.

Seasonal rates are a time-of-use rate that vary according to the season. For example, a summer-peaking utility will usually charge a higher demand and/or energy charge for the summer months than for the winter months. In this case, the analyst computes annual savings by separating the months the client is on each rate, calculating each season, and then totaling the two seasons.

Whatever rate a client is on, the analyst needs to calculate savings the same way that electric bills are computed, not with average rates that distort the savings message.

### Economic Decision Statistics

Two economic decision statistics that are used in lighting evaluations will be discussed here. The most common statistic used is the simple payback (SPB), but this measure has drawbacks for long-term projects. Some corporations and the federal government use life-cycle costing (LCC) for the evaluations of return on investment for their lighting upgrades or relighting projects.

#### Simple Payback

SPB relates to how long it takes to recover an initial investment in a cost-saving measure, assuming the annual savings remain constant and the time value of money is unimportant. To calculate simple payback, simply divide the initial investment by the annual savings. This calculation is shown in Chapter 8, Fact Sheet 13.

SPB suffers from two key flaws. First, it assumes that $100 received one year from today is equivalent to $100 received 5 years from today, yet most business organizations assign a higher value to dollars received sooner than those received later. Second, SPB does not consider the effects of different lives of the alternatives considered. For example, Projects X and Y each cost $10,000 and save $2000 each year; therefore, both have a SPB of 5 years, making them appear equally acceptable. However, if Project X has a useful life of 5 years and Project Y has a useful life of 10 years, Project Y is obviously a better investment.

When discount rates are low and the useful lives of all the alternatives are the same (which they often are for lighting upgrades), SPB becomes more accurate. Despite the flaws outlined above with SPB, most organizations use it in making capital budgeting decisions, preferring the intuitive simplicity of the statistic. However, SPB should only be used as a guide for making accept/reject decisions and not as a ranking statistic or a means of making system selections among a group of possible alternatives. Life-cycle costing techniques, discussed in the following section, overcome the limitations of SPB and provide for better decision making in these situations.

#### Life-Cycle Costing

LCC is an economic method of project evaluation that takes into account all costs arising from the ownership, operation, maintenance, and disposal of a lighting project. Instead of focusing on the short-term time to pay back that SPB does, LCC considers long-term economic performance. Life cycle cost is the total cost of owning, operating, maintaining, and disposing of a lighting system over a given period of time with all costs adjusted (discounted) to reflect the time value of money. When faced with the decision to choose a single option among several, any of which meet the same need (mutually exclusive alternatives), the lowest LCC will always provide the economically optimum decision.

## Auditor Evaluation Skills

A good auditor should be a perceptive observer, with good investigative skills, honed by performing many audits. A perceptive auditor is aware of the lighted environment and understands what is being observed directly through the sense of sight. This skill requires the ability to look at lighting systems and the spaces they light in a particular way with regard to certain characteristics and the ability to inventory and categorize equipment.

Perception also requires experience, which develops the intellectual capacity to correctly assess lighting conditions. This ability is sometimes called insight or discernment. An awareness of certain lighting conditions needs to be developed so that an auditor eventually gains knowledge of lighting systems and the lighted environment through his or her own perceptions. Since our perceptions of the same scene can often differ from others' perceptions, it is good for teams of two or three auditors to compare their impressions. Practice in performing audits makes auditors alert to spot key features of a space and its lighting and fine-tunes their quickness of response. A seasoned auditor will not have to spend a lot of time in each space.

An auditor becomes skilled in the process of investigation, not unlike a good detective. Investigative skills involve a search for the facts or evidence using a detailed inquiry or systematic examination to uncover and report information that may be hidden from others.

Auditors should also be skilled in interview techniques because some of the critical information for an audit must be obtained from interviews with key building personnel. Listening to building management personnel and occupants and sorting conflicting information are also helpful skills. Interviews are usually conducted face-to-face and one-on-one in a conversational way that elicits relevant information without making subjects feel as if they are being interrogated. Nonthreatening places such as coffee break rooms are good settings for these interviews.

Sometimes auditors can know too much! It is better for audit team members to have great observation and counting skills rather than detailed technical knowledge of lighting. When conducting an audit or survey, lighting engineers and designers with a high level of technical knowledge can get too involved with how they would change the existing lighting systems or the space, and this preoccupation can obscure the primary mission of the lighting audit, causing data collection to be irregular. Later in the audit process, during the analysis phase, is when a lighting professional should analyze the existing lighting systems and consider upgrade alternatives.

## Selection and Training of Auditors

Auditors are recruited based on their possessing the observation and investigative skills described in the previous section. In addition, stamina is an important attribute for auditors. The athletic endurance to be able to work long hours without sitting down is probably the most important ability.

Once they are selected, auditors are trained in a classroom setting, receiving instruction on lighting basics and other lighting facts (see the Lighting Fact Sheets in Chapter 8). Immediately following classroom training, guided field surveys and audits are conducted and discussed on-site. After formal training, the auditor's observation skills are tested and honed through experience and on-the-job training. There are workshops and seminars to help train auditors in hands-on sessions.

## Summary

The evaluation of lighting systems is a multistep examination process, consisting of at least three steps: (1) the lighting survey or audit, (2) identification of opportunities for improvement, and (3) calculation of savings and payback. Since Steps 2 and 3 are significantly affected by Step 1, surveys and audits should be performed with diligence to avoid mistakes and problems.

Potential improvement opportunities may be identified on a quantitative or qualitative basis. Quantitatively, improvements can be measured and compared with standards of performance on the basis of power density (W/sf), illuminance (FC), and controls. Qualitatively evaluating improvements is a more subtle and highly subjective task, made easier by auditor experience. Qualitative lighting attributes include veiling reflections, glare, color and CRI, lamp flicker, luminance ratios, and visual comfort probability.

Savings may be calculated either as simple payback (SPB) or using life-cycle costing (LCC). SPB should only be used as a guide for making accept/reject decisions and not as a ranking statistic or to make system selections among a group of alternatives. LCC techniques overcome the limitations of SPB and provide for better decision making in these situations. Training and experience provide auditors with the skills they need to be good investigators and perceptive observers.

# Chapter 3
# Types of Surveys and Audits

This chapter outlines the characteristics and technical details of each of three types of surveys and audits: the walk-thru survey, the intermediate audit (mini-audit), and the comprehensive audit. The information in this chapter will help the auditor determine which type of audit best suits the needs of a particular project. In addition, sufficient details are presented to enable more accurate and efficient surveys and audits for both beginning and experienced auditors. The chapter also includes information concerning how to perform data analysis and write reports for each type of survey.

## Which Type of Survey or Audit Should I Use?

The type of survey or audit to use will be determined by several factors. The purpose, reason, or need for performing a survey or audit will help determine which type is required for each project. For example, if the building owner or property manager simply wants a "checkup" of the lighting systems or wants to know if there is potential for a lighting upgrade that will pay back in a certain time, then a walk-thru survey will satisfy these needs. However, if systems have been in place for more than 20 years and not much has been done to upgrade to newer technology, then the comprehensive audit is more suitable. The type of building can also help determine the type of survey or audit. For example, many houses of worship have been surveyed with a walk-thru. Limited resources, in these cases, influenced the selection of the walk-thru. In other words, the cost of the survey or audit can also be a factor in deciding which type of survey to use. The size and scope of a facility, along with the number of different lighting systems in it, also inform the decision. Intermediate audits are commonly used for schools, hospitals, hotels, and restaurants. However, a college campus with many different building types, built over several decades, will have many different types and vintages of lighting systems and will benefit most from a comprehensive audit. Also, when the client/customer wants to explore several upgrade or relighting options, the comprehensive audit is most suitable. Table 3.1 summarizes the selection criteria.

The more data required, the more measurements needed—and more measurements require more analysis. As the need for more data and measurements becomes apparent, the auditor is moved up the hierarchy from surveys to audits. Firms that perform audits on a regular basis have little difficulty deciding which type of survey or audit is best for any given project. This decision will be based on their past experience involving many different building sizes and types, the scope of the upgrade, and the cost of the survey or audit.

## The Walk-Thru Survey

The simplest type of survey/audit is the walk-thru survey. In the walk-thru, the intent is to collect just enough information to make effective recommendations in the shortest time. The lowest level walk-thru

**Table 3.1** Selection Criteria for Determining Appropriate Type of Survey/Audit

| Type of Building, Need, or Cost | Walk-Thru Survey | Intermediate Audit | Comprehensive Audit |
|---|---|---|---|
| Checkup | X | | |
| Upgrade Potential | X | | |
| Houses of Worship | X | | |
| Low Cost | X | | |
| Offices | | X | |
| Schools | | X | |
| Hospitals | | X | |
| Hotels or Restaurants | | X | |
| Medium Cost | | X | |
| Older Lighting Systems That Have Not Been Upgraded | | | X |
| Relighting Projects | | | X |
| College Campuses | | | X |
| Manufacturing Facilities | | | X |
| High Cost | | | X |

survey would be done to gather the most meaningful facts needed for a one-for-one retrofit. For the most basic form of walk-thru surveys, no fixtures are counted and no calculation of power density is made. The surveyor briskly tours the building in search of simple energy conservation measures (ECMs). (Chapter 9 describes a number of possible lighting ECMs.) In this basic survey, information should be provided on a per-fixture basis. Only fixture information is required in this case, and the report of opportunities for improvement is limited, usually to one upgrade option for each fixture type. In fact, if a written report is not needed, this information may be provided while the surveyor is still on the client's premises! The client can easily determine the total cost and savings from making the recommended upgrade by multiplying the per-fixture savings by the number of fixtures of that type in the building.

When a walk-thru survey is performed as a "checkup" of the existing lighting systems, a report card is provided to the client/customer. This report should be limited to the condition of the lighting systems examined and ideas for upgrading these systems to save energy, improve working conditions, and reduce maintenance costs. Any questions that the building manager, engineer, or maintenance personnel may want answered about lighting problems or concerning retrofit products under consideration can be answered during the time of the walk-thru. A recommendation often made in checkup reports is to perform a more thorough and in-depth audit to reveal a higher level of improvements and savings. This recommendation should only be made if the checkup reveals the potential for such improvements and savings that will be cost-effective.

One of the tasks of the walk-thru may be verifying the general accuracy of owner-furnished architectural, mechanical, and electrical plans and drawings. This is best done by a two-person team. The question that the survey team needs to ask in each space is, are the lighting systems we are looking at the same as those that are shown on the plans? If they are not, then perhaps an intermediate or comprehensive audit is more appropriate so that physical dimensions can be measured. Sometimes the firm performing the audit will use the survey to make an estimate of further audit work and may not be paid for the walk-thru.

## "Free" (Vendor) Surveys

Vendor surveys are usually the walk-thru type, done at no cost to obtain retrofit work. Energy services companies (ESCOs), lighting management companies, and others who provide lighting upgrades may provide the survey at no cost, but the value of such a walk-thru may be low, if what is needed is a more extensive audit. On the other hand, if a simple upgrade to T8 lamps and electronic ballasts is the only requirement, then the free walk-thru survey may be appropriate.

In many cases the walk-thru survey is done with a particular luminaire upgrade in mind (see Chapter 8, Fact Sheet 9). With this approach, some well-established wisdom applies: "When all you have is a hammer, everything looks like a nail." This is usually the type of free survey that is done by vendors of a particular lighting retrofit product.

Be aware that free surveys may be worth exactly what they cost! For example, a lighting retrofit company that specializes in converting fluorescent lighting systems from T12 to T8 lamps may only gather information about the number of fixtures and the number of lamps in each fixture, because the simplest way to do this upgrade is a one-for-one replacement. While this simplistic approach saves energy, it falls short of providing the lasting benefits that could be realized if a more comprehensive approach were taken. If a simplified approach is what the building owners believe they need, then the surveyor should report that this approach may not have the desired long-range results.

Effective Lighting Solutions, Inc. (the author's company) has performed comprehensive audits in spaces that have originally been surveyed by vendors. In these cases, it was found that the free vendor survey resulted in a serious undercount of the quantity of fixtures. For building operators that are trying to budget a lighting upgrade, fixture miscounts will cause difficulties in estimating the cost of the upgrade. The quantity of fixtures is the key multiplier in the upgrade cost calculation, and any error in the count throws the cost of the project off. If the fixture count is underestimated, a project that initially seems affordable, may exceed a capital expenditure limit when the actual count is discovered.

The risk of providing an inadequate lighted working environment is high using vendor surveys. Light levels may end up too high or too low, because illuminance (light level) measurements are not made in a walk-thru survey. Illuminance measurements are required when values are needed to compare with recommendations for current tasks. This would usually be done in an intermediate or comprehensive audit to prepare for an upgrade involving several technologies or for an extensive relighting. In addition, when the "apparent brightness" of the T8 lamp is not accounted for, many end users complain that there is too much light (see Chapter 8, Fact Sheet 1).

## Information Gathered for the Walk-Thru Survey

In the walk-thru survey, rather than record detailed information, as is done for audits, auditors record visual impressions of the working environment, the lighting systems, and the condition of the facility, sometimes using a tape recorder. There is no attention paid to the lighting needs of workers or their visual tasks at this level of survey. Estimates rather than measurements are used in the walk-thru survey. Room size, equipment inventory, lighting load, and hours of use are all estimated. Architectural drawings may be useful for room size data—if the drawings have been kept up-to-date.

It is helpful to calculate a power density figure for the major spaces being surveyed. Power density (in watts per square foot) is the best indicator of lighting system efficiency and can be used to screen potential upgrade or relighting opportunities. To calculate power density, the lighting load for a space and the area of the space are needed. This calculation is detailed in Chapter 9, Fact Sheet 13.

The lighting load depends on the quantity and type of fixtures in the space. The quantity of fixtures may be counted or found on building drawings. Fixture quantities may also be estimated on a density basis, especially if standard layout patterns are used throughout the facility (see Chapter 8, Fact Sheet 8). Other fixture data may be obtained from observation or taken off the current building drawings. Fluorescent and HID systems require information regarding the ballasts that are in the fixtures. In the walk-thru survey, this information is estimated from ballast manufacturers' catalogs or other sources such as the Green Lights *Lighting Upgrade Manual.* Information on the type of ballast and the number of lamps powered by each ballast is needed to look up values of "input watts" or "watts per fixture." The lighting load (in watts) for the system lighting a particular space is found by multiplying the number of fixtures times the watts per fixture.

Hours-of-use values are the most difficult to estimate. For the walk-thru survey, it is best to get this information from the building operator or engineer. Building personnel may provide this information in the form of building operating schedules. Hours of use are then calculated from these building schedule estimates. One source for this information is the DOE/EIA (Department of Energy/Energy Information Agency) publication *Lighting in Commercial Buildings.* The data in this publication is derived from building census information for various types of buildings. Sample data is shown in Table 3.2.

**TABLE 3.2** Sample Hours-of-Use Data for Various Building Types

| Building Type | Annual Hours of Operation |
|---|---|
| Assembly | 2760 |
| Education | 2605 |
| Food sales | 5200 |
| Food service | 4580 |
| Health care | 7630 |
| Lodging | 8025 |
| Mercantile | 3325 |
| Office | 2730 |
| Warehouse | 3295 |
| Average nonresidential | 3500 |

Careful thought and sometimes a little sleuthing are needed to obtain reasonably accurate hours-of-use information. For example, do not overlook spaces such as corridors and stairwells, where there may be no switches to turn the lights off and they burn constantly, 8760 hours a year (in leap years, 8784 hours). It is common, but not wise, to forget to include off-normal conditions such as the occasions when nighttime cleaning crews turn on all the lights but do not turn them off when they leave an area. A midnight drive by the building may help get a true estimate of the hours of use.

Light levels may be estimated from observation of the appearance and impressions of the brightness of the spaces. In many situations, it does not take a light meter to tell if a space has too much or too little light. A closer look may reveal that the reason for too little light is a deterioration of room surfaces or a lack of lighting maintenance. As shown in Chapter 8, Fact Sheet 14, light levels can deteriorate as much as 60% for neglected systems. Lamps and ballasts need to be replaced at end of life to maintain light levels.

When this condition is encountered, the surveyor has a difficult choice to make. If the deteriorated condition is taken as the base condition, upgrades will not usually be cost-effective. The client/customer needs to be informed of this condition, and the base should be set at the level for which the system was originally designed. The only situation where this has not been acceptable is when working with performance contractors who must show a reduction in electric bills in order to satisfy their contract and get paid.

Renewing deteriorated room surfaces will increase light levels and provide a more attractive work environment. The problem is that most people, unfamiliar with lighting, do not understand the connection between lighting and room surface materials, colors, and textures. The surveyor needs to make this connection for the building owner/manager by simply explaining that the more light that is reflected onto the work plane, the less light is needed to start with, so lighting systems are more effective and efficient in interiors with light-colored surfaces.

High light levels waste energy and, what is worse, degrade lighting quality. High levels of light are often accompanied by glare situations, which are detrimental to worker productivity. Interviews of workers in these environments will often reveal dissatisfaction and a probable cause of expensive absenteeism.

Higher-level (slightly more complex) walk-thru surveys will need more data to accomplish their purposes. For example, if power density is being used as a screening measure, then room dimensions and lighting loads are needed to obtain watts per square foot (W/sf). For a walk-thru, this information is usually estimated or taken off the architectural drawings. If potential savings are being calculated, then hours-of-use information will be required. Lighting system voltages will be needed for retrofit equipment ratings and is usually obtained from an interview or taken off the electrical drawings. Light level and switching/zoning data are not usually collected during walk-thru surveys.

### Data Analysis

The data analysis for a walk-thru survey is simple and straightforward. Power density may be calculated for the building or for selected spaces, based on estimates of lighting loads and space areas. Some surveyors provide a spot check of power density for selected areas. For the simplest of walk-thru surveys, power density may not be calculated at all. At this most basic level, fixture counts are not made and calculations of savings are usually limited to energy savings only and are performed on a per fixture basis.

## Example of Per Fixture Savings and Payback

Spaces with 2 × 4, 4-lamp fixtures equipped with 40-watt T12 lamps and older coil and core magnetic ballasts are calculated in the following way: In the Green Lights *Lighting Upgrade Manual,* fixture 717 is listed at 190 watts. A one-for-one upgrade of the fixture to T8 lamps and one 4-lamp electronic ballast is recommended. The Green Lights *Lighting Upgrade Manual* lists fixture 729 (which has a low ballast factor, or low BF) at 95 watts. The per fixture savings: 190 watts - 95 watts = 95 watts. If estimates of hours of use are used, then the lighting energy savings can be calculated. For this example, if the operation of these fixtures is full-time, then the energy saved per position is: 95 watts/1000 × 8760 hours = 832 kilowatt-hours (kWh). The energy cost savings is found by multiplying the energy savings times the average cost per kWh. If the average cost is $0.06, the annual cost savings are: 832 kWh × $0.06/kWh = $50.

The upgrade cost is found by adding the material cost of the new lamps and ballast to the labor cost to install the upgrade. For the example, the four T8 lamps cost $2 each and the electronic ballast costs $20. The material cost is: 4 × $2 + $20 = $28 per fixture. If it takes the lighting management firm an average of 3/4 hours per fixture, and their labor rate is $60/hour, then the labor cost is: 0.75 hours/fixture × $60/hour = $45 per fixture. The total installation cost (per fixture) is: $28 + $45 = $73. The simple payback (SPB) for this upgrade is calculated by dividing the total cost of each fixture upgrade by the annual savings per fixture. To complete the example, the SPB per fixture = $73/$50 per year = 1.5 years, or 18 months.

When fixtures are counted, simple payback calculations are made from estimates of upgrade costs and savings. Savings calculations may be limited to energy savings only, or they may include maintenance savings. Energy savings are calculated from estimates of lighting loads and hours of use.

## Example of Energy Savings and Payback

The lighting load for a room with nine 2 × 4, 4-lamp fixtures equipped with 40-watt T12 lamps and older coil and core magnetic ballasts is calculated in the following way: In the Green Lights *Lighting Upgrade Manual,* fixture 717 is listed at 190 watts. Calculate: 9 fixtures × 190 watts/fixture = 1710 watts; 1710 watts × 1/1000 = 1.71 kW.

The building personnel estimate that this space is used for 5 days, 12 hours per day, 50 weeks per year. Hours of use is calculated as: 5 days/week × 12 hours/day × 50 weeks/year = 3000 hours/year. Lighting energy is found by multiplying the lighting load times the hours of use. Thus, the lighting load of 1.71 kW is multiplied times the hours of use of 3000 hours. This calculation (1.71 kW × 3000 hours = 5130 kWh) provides the present (before-upgrade) annual lighting energy for this space.

If an average electric rate is used, which is common for walk-thru surveys, then the economic savings can be found by multiplying the kWh savings times the average cents/kWh rate. If the average rate is $0.08/kWh, the lighting energy cost for this space is: 5130 kWh/year × $0.08/kWh = $410.40 per year.

Calculation of the upgrade lighting load, lighting energy, and cost is performed in the same manner. For an upgrade to nine new 3-lamp T8 parabolic luminaires, the new lighting load is calculated in the following way: In the Green Lights *Lighting Upgrade Manual,* fixture 569 is listed at 91 watts. Calculate: 9 fixtures × 91 watts/fixture = 819 watts. If the same schedule is anticipated after the upgrade, and if no controls are added, then the annual hours of use will still be 3000 hours. Thus, the lighting energy is: 819 watts/1000 × 3000 hours = 2457 kWh. The lighting energy cost after the upgrade will be: 2457 kWh × $0.08/kWh = $196.56.

Energy savings for this upgrade to new luminaires is found by subtracting the new lighting energy value from the present value: 5130 kWh/year - 2457 kWh/year = 2673 kWh/year. Annual lighting energy cost is found by subtracting the new lighting energy cost from the present cost: $410.40 - $196.56 = $213.84.

For upgrades to new lamps, there will be a difference in lamp (bulb) costs. Lamps using newer technology usually cost more than the bulbs in current use. The difference should be subtracted from the energy and maintenance savings. In order to calculate annual lamp and labor cost, the average number of lamps replaced per year needs to be calculated. A standardized way of calculating the average number of lamps used in a year is shown in Equation 3.1 and is repeated in Chapter 8, Fact Sheet 13.

**COST REDUCTION CALCULATION FOR LIGHTING UPGRADE**
**LOCATION: XYZ Company**
**Replace 2x4 lensed troffers w/new luminaires**

| | Before | After | Difference |
|---|---|---|---|
| LAMP ID | F40T12CW | FO32/741 | |
| LAMP WATTS | 40 | 32 | |
| LAMPS per FIXTURE | 4 | 3 | |
| LAMP LIFE | 20,000 | 20,000 | |
| #FIXTURES | 9 | 9 | |
| WATTS per FIXTURE | 190.0 | 91.0 | |
| LIGHTING LOAD (kW) | 1.71 | 0.819 | 0.89 |
| OPERATE HRS per YR | 3,000 | 3,000 | |
| KWH per YEAR | 5,130 | 2,457 | 2,673 |
| CO$T per KWH | $0.080 | $0.080 | |
| ANNUAL ENERGY CO$T | $410.40 | $196.56 | $213.84 |
| | | | |
| LAMPS per YEAR | 5 | 4 | |
| CO$T per LAMP | $1.20 | $2.00 | |
| ANNUAL LAMP CO$T | $6.48 | $8.10 | ($1.62) |
| | | | |
| HRS to SPOT REPLACE | 0.50 | 0.50 | |
| $LABOR/HR | $25.00 | $25.00 | |
| ANNUAL LABOR CO$T | $67.50 | $50.63 | $16.87 |
| TOTAL ANNUAL CO$T$ | $484.38 | $255.29 | $229.10 |
| ANNUAL SAVING$ | | $229.09 | |
| | | | |
| CO$T of LUMINAIRES | | $787.50 | |
| CO$T of LAMPS | | $54.00 | |
| CO$T of INSTALLATION | | $674.50 | |
| TOTAL CO$T of UPGRADE | | $1,516.00 | |
| SPB = | | 6.6 | yrs |

**FIGURE 3.1** Spreadsheet showing cost reduction calculations for lighting upgrade example.

$$\text{Lamps per year} = \frac{\text{\# lamps / fixture} \times \text{burn hours / year}}{\text{average rated lamp life}} \quad (3.1)$$

## Example of Lamps and Labor Cost Savings and Payback

Using the equation for lamps per year and the data from the previous example: 3 lamps/fixture × 9 fixtures × 3000 hours per year/20,000 hours = 5 lamps/year. (The average rated lamp life in this example is 20,000.) In the spreadsheet shown in Figure 3.1, five T12 lamps ("before") and four T8 lamps ("after") are replaced each year for this system.

Lamp costs are calculated by multiplying the average number of lamps per year by the cost per lamp. For the example, annual before-upgrade lamp costs are calculated by multiplying the average number of lamps replaced per year times the cost per lamp: 5.4 × $1.20 = $6.48. The difference in lamp costs is calculated by subtracting the new lamp cost from the old lamp cost: $8.10 - $6.48 = ($1.62).

Labor to replace lamps is calculated by multiplying the average number of lamps per year times the estimated hours to spot replace each lamp times the labor rate. For this example: 5.4 lamps × 0.5 hours × $25/hour = $67.50 per lamp for the "before" case. For the "after" case, 4.1 lamps × 0.5 hours × $25/hour = $50.63. The labor savings is found by subtracting the "after" labor cost from the "before" labor cost: $67.50 - $50.63 = $16.87. (This savings results from fewer number of lamps in the system, since the old fixtures had 4 lamps each and the new fixtures have 3 lamps each.)

## Example of Total Cost Savings

The total annual costs are calculated by totaling the annual energy cost, lamp cost, and labor cost. For the "before" case: $410.40 + $6.48 + $67.50 = $484.38. For the "after" case, $196.56 + $8.10 + $50.63 = $255.29. Cost savings are calculated by subtracting the "after" total annual costs from the "before" total annual costs: $484.38 - $255.29 = $229.09.

## Example of Upgrade Cost

Cost of the upgrade is found by multiplying the number of luminaires times the cost (9 luminaires × $87.50 = $787.50) and adding the lamp cost. Each luminaire uses 3 lamps that cost $2 each. Total cost of lamps is 9 × 3 × $2 = $54. Total material costs are:

\$787.50 + \$54 = \$841.50. Installation costs are calculated by multiplying the labor cost per luminaire times the number of luminaires or by using a labor rate per luminaire times the hourly rate. For our case, the successful bid from an electrical contractor to remove the old fixtures and install the new fixtures that we purchased was \$674.50. Total installation cost is found by adding the material cost and the labor cost: \$841.50 + \$674.50 = \$1516.

Example of simple payback is calculated once the costs and savings are computed in Equation 3.2.

$$SPB = \frac{\text{cost of upgrade}}{\text{savings / year}} = \frac{\$1516}{\$229.09} = 6.6 \text{ years} \tag{3.2}$$

Most surveyors do not include the effect of the lighting upgrade on the building HVAC systems at this level. These interactions may be included in the comprehensive audit.

Reporting the results of the walk-thru survey and the analysis of the information is simply a matter of recording in a brief format what types of fixtures and lamps are in use, the condition of the systems, the recommendations for improvement, the estimated cost of the improvements, the associated savings, and a simple payback. If spot checks were made of power density, then these are reported and compared with the potential "after" improvement.

## The Intermediate Audit (Mini-Audit)

The intermediate audit (also called the mini-audit) is usually a hybrid of a walk-thru survey and comprehensive audit; that is, it contains elements of each. It is more detailed than a walk-thru survey, and the data collection is usually done by a small team of two or three.

If room dimensions are available from reliable building drawings, they will be used. If they are not, then each room should be measured individually. Room measurements can be facilitated by using the measuring tools described in Chapter 4.

Power density for major areas should be calculated from room dimension data and from lighting loads that are either estimated or measured. Instead of using one estimated average occupancy schedule for the building, occupancy schedules should be determined for each major area. This hours-of-use information may be obtained either by interview or by measuring with lighting loggers. (See Chapter 4). Lighting equipment inventories should either be estimated or fully counted, depending on the accuracy requirements of the audit.

The auditors should be looking for upgrade opportunities based on equipment type, and the focus is still primarily on replacing individual components on a one-for-one basis. This level of audit is sometimes concerned with the exact mix of lamp types and ballasts, depending on the degree of accuracy required. The intermediate audit is the level where auditors start to record their subjective judgments of the visual tasks performed, the state of maintenance, the surface finishes, and the overall lighting quality. Lighting maintenance routines are also assessed at this level of audit. The Green Lights survey may be considered an intermediate audit.

Hospitals and elder care facilities, schools, and many offices are audited using the intermediate audit since these facilities are characterized by numerous fixtures of the same type, laid out in repeated patterns. When performing audits for hospitals or other health care facilities, the lighting professional leading the audit should be aware of the IESNA Recommended Practice publication *Lighting for Hospitals and Health Care Facilities* (RP-29-95). The intermediate audit is also used for hotels and restaurants.

The financial analysis made for the intermediate audit is usually more than SPB, but it still may not be a full-scale life-cycle costing procedure. Calculations are usually performed for the complete project instead of on a fixture or component basis. A rate-of-return calculation may be performed, such as that provided by the Quikalc or Sylvia PC computer programs described in Chapter 6.

Reports usually include equipment inventories, power densities, limited evaluation of upgrade or relighting alternatives, and payback. The report will compare the existing lighting systems to their potential efficiency.

## The Comprehensive Audit

The comprehensive audit is the most complex. The object is to reveal three major facts about the lighted environment, listed in priority order:

1. What visual tasks are being performed?
2. Are the lighting systems performing effectively to light those spaces and tasks?
3. How are the lighting systems being operated and maintained?

The needs of the occupants must not be neglected or given low priority because people are the very reason that building lighting systems have been provided — so that workers may perform productively.

This high-level audit must not only provide a physical inventory of the lighting equipment but also must evaluate the equipment performance and provide an accurate assessment of the visual tasks performed. To provide this extensive information requires an evaluation of each working space. This means that the comprehensive audit will go far beyond the simple inventory of lighting hardware made in walk-thru surveys, and the two- or three-person audit team will spend considerably more time in each space. In this broad, systems-oriented approach, all aspects of the lighted environment should be considered, including visual tasks and lighting quality. The focus is on the needs of the occupants, so auditors should record their subjective reactions to each space and include interviews of occupants and their evaluations of the spaces in which they work.

A comprehensive audit is performed when the highest level of detail is required. The need for collecting thorough equipment details is to provide multiple upgrade options or a relighting plan. To provide the detailed information required, all lighting equipment must be fully identified and counted. Attention will be paid to the actual mix of lamps and ballasts in use. Making detailed electrical load measurements will alleviate the need for detailed visual inspections.

Illuminance measurements should be made in all spaces and at specific tasks. This means that most measurements are made at a desktop or benchtop. Corridors are measured at floor level. A digital light meter is recommended for footcandle (FC) measurements. Light meters are discussed in Chapter 4, and the techniques of measuring illuminance are described in Chapter 5. Several readings need to be taken in each space, and any obvious underlighted or overlighted areas must be measured and recorded separately.

Daylight contributions are accounted for by first measuring with the electric lighting on and then measuring with the lighting off. See Chapter 5 for details about measuring daylight contributions. The assessment of daylighting potential is optional, depending on the purpose and scope of the audit.

Fixture spacing data is recorded for spacing calculations and to analyze the effects of particular luminaire efficiency improvements such as a reflector upgrade or a change from lens to louver. Chapter 8, Fact Sheets 5 and 13 describe spacing in detail and prescribe how to perform the calculation.

The comprehensive audit should also account for the facility's total lighting energy consumption and electrical demand. A majority of the building lighting energy needs to be accounted for. The measurement of lighting system voltages needs to be made at the lighting distribution (circuit-breaker) panels. Load measurements of branch circuits should be made at these panels using a clamp-on ammeter or power meter. With this instrumentation it is not necessary to know the exact mix of lamp types and ballast vintages, since these measurements will reflect the mixed loads. These measurements should be made by a qualified electrician using considerable care for electrical safety. Since opening the lighting distribution panels poses some risk of interrupting business operations, they should be scheduled and performed with extreme caution. Chapter 4 provides details of audit instrumentation.

Power quality (PQ) may be measured to identify harmonics, transients, or other disturbances that may affect the operation of the lighting circuits. Attention to neutral wiring and testing the integrity of the grounding wiring will provide assurance that the electrical distribution can operate electronic lighting equipment without faults. A power quality specialist may be contracted to join the audit team or perform a separate PQ audit.

Lighting loggers should be placed in representative spaces for extended observation periods to obtain accurate hours-of-use data. (Loggers are described in Chapter 4 and their use is outlined in Chapter 5.) An observation period of 1–2 weeks is recommended. An alternative to loggers is a kWh meter or data collection system that monitors the lighting circuits for the same period of time. Strip-chart recorders are not recommended for these measurements because they are cumbersome to use and the volumes of charts produced are time-consuming to analyze.

The audit team should look for opportunities to employ control strategies to control the hours of use and/or the lighting power for each space. Controls can reduce the electrical energy used in lighting systems in two ways. Electrical energy (kWh) can be reduced either by reducing the lighting power (kW) or by reducing the hours of use (time), and since electricity is paid for per kWh, reducing either will reduce electric costs. Consideration should be given to switching, occupancy recognition, scheduling, lumen depreciation compensation, and daylight harvesting. Several of the lighting energy conservation measures (ECMs) detailed in Chapter 9 describe specific lighting controls.

## Zonal-Cavity Calculations

An extensive analysis should be performed including calculations of power density, zonal-cavity calculations, and luminaire spacing. Part of the analysis will be to compare existing lighting systems with the current recommendations found in the latest IESNA Recommended Practices (RP). These calculations are outlined in Chapter 8, Fact Sheet 13. The zonal-cavity calculation is usually used for new or relighting projects, but the use of this calculation for upgrade purposes is helpful for comparing the field calculation with the "actual" coefficient of utilization (CU) of a system. When used in this way, Equations

8.5 and 8.7 (Chapter 8) are combined and rearranged to solve as equation 3.3 for CU as shown here:

$$CU = \frac{FC\,(\text{maintained}) \times \text{length} \times \text{width}}{\#\text{ of fixtures} \times \#\text{ of lamps/fixture} \times \text{lumens/lamp} \times LLF} \tag{3.3}$$

(*Note*: The value for lumens/per lamp comes from the lamp manufacturer's catalog.)

## Example of Zonal-Cavity Calculation to Find CU

A corporate dining room has a luminous ceiling consisting of 4-foot, 2-lamp strip lights equipped with 34-watt reduced-wattage fluorescent lamps. There are 120 fixtures in an area 30 feet by 40 feet (1200 square feet). The power density is (approximately) 7–8 W/sf. The shielding media covering the entire ceiling is an "egg crate" latticework metal louver with openings of approximately 1-inch square. This metal louver blocks a lot of the light from the strip fixtures, mounted about 2 feet above the louver. In addition, there are about 15% burnouts and the lamps appear old. There is dirt on the sides of the louver openings. An average of 50 FC is measured at the dining room tables, located 2.5 feet from the floor.

Because the lamps are observed to be old and there is a high potential for both "energy-saving" (reduced-wattage) lamps and "energy-saving" ballasts, a lower ballast factor must be considered. In addition, dirt on the louvers lowers the light loss factor (LLF) to an estimated 55%. The lumens/ lamp value from the manufacturer's catalog is 2825. Using Equation 3.3:

$$CU = \frac{FC\,(\text{maintained}) \times \text{length} \times \text{width}}{\#\text{ of fixtures} \times \#\text{ of lamps/fixture} \times \text{lumens/lamp} \times LLF}$$

$$= \frac{50 \times 30 \times 40}{120 \times 2 \times 2825 \times 0.55} = \frac{60{,}000}{372{,}900} = 0.16$$

Using this calculation is more accurate than the auditor's original CU estimate of 0.25, which was about 50% too high. (A faster, more accurate way to calculate the CU using an iterative method is outlined in Chapter 6, using this same example, in the section discussing Calcu-Lite software.)

Comparison of the existing luminaire centers with the maximum spacing calculated using the spacing criteria for the luminaire will provide valuable information regarding the light distribution in the space and the effect of changing the distribution with luminaire upgrades such as reflectors and louvers. This calculation is detailed in Chapter 8, Fact Sheet 13.

### Maintenance Costs

A maintenance cost analysis should be performed using the data collected in order to minimize the facility lighting maintenance cost. Most facilities use a large number of different types of lamps and ballasts. A common recommended improvement is to consolidate and reduce the number of different types of lamps. For example, if a building uses 4-foot lamps in the office and 8-foot lamps in its manufacturing spaces, upgrades using 4-foot T8 lamps only should be given priority over combinations of 4-foot and 8-foot lamps, to cut down lamp inventory. There will be cost savings associated with this recommendation that may be calculated and presented in the report.

Another recommendation commonly made in the maintenance cost area is to recommend group maintenance practices. This improvement will keep light levels from depreciating to unacceptable levels by maintaining lighting system performance. As shown in Chapter 8, Fact Sheet 14, neglected lighting systems lose efficiency over time. Periodic, scheduled lighting maintenance saves money because it calls for preventive maintenance on large groups of luminaires instead of relying on reactive "breakdown" maintenance when individual lighting components fail. Group relamping can lower labor costs and avoid lost productivity from work interruptions when spot relamping is done during normal work hours. Note that when lamp changers relamp office areas during working hours, they usually interrupt workers near the luminaires being relamped and those close by who may join the conversations that are initiated by the friendly lamp changers. The benefits of changing all the lamps in a group at scheduled intervals has the additional advantage of lowering inventory costs because there are fewer lamps to stock. A cost analysis should be performed showing the potential labor and inventory savings and this analysis should be presented in the report.

Relamping intervals can be calculated as shown in Chapter 8, Fact Sheet 13. There is, however, a simple technique that does not require calculations, which some maintenance organizations use to determine when it is time to group relamp. Most companies that practice group relamping also replace early burnouts on a spot-replacement basis between major relampings. If they have determined they would normally allow 10% burnouts before major relamping, they simply purchase 10% more lamps at the time of group relamping. When this extra supply of lamps is depleted by spot relamping, it is time to group relamp again.

A comprehensive analysis will also include a detailed electric rate analysis and life-cycle costing. Average rates calculated by using the total cost of the electric bills for a year for a facility and dividing by the total kWh for the year are not satisfactory for many reasons. Electric bill savings should be computed the same as electric bills are paid. Using "average rates" can distort the economic savings when there are demand charges, seasonal rates, and declining block rates (see Chapter 2). Electric utilities can help the analyst with these calculations and often will provide calculation worksheets with the appropriate multiplying factors filled in. In working with the rates, one service that some analysts perform is to inquire of the utility if their client is on the most economical rate, or if there are potential savings from consolidating multiple electrical service entrances to a building.

As explained in Chapter 2, life-cycle costing is the sum of time-equivalent costs of acquiring, owning, operating, and maintaining lighting systems over a designated period of time. The principles of life-cycle costing require that lighting alternatives must be equivalent in their function and performance, that all significant initial and future costs must be considered, and that costs must account for the time value of money. Initial costs include equipment, installation, wiring, and HVAC costs. Energy costs, including the direct cost of lighting and lighting-related HVAC costs, are the dominant component of life-cycle cost for lighting systems. Maintenance costs are those associated with maintaining the lighting system over its lifetime. Energy and maintenance costs are sometimes grouped together and called annual operating or running costs.

## Example of Life-Cycle Costing

An existing system (A) and a lighting upgrade (B) are to be compared. System A has an initial cost of $40. Upgrade B has an initial cost of $80. We want to calculate the lifetime owning and operating cost, LOC. The life of the lighting system and the upgrade is 10 years.

System A consists of one 2 × 4, 4-lamp luminaire rated at 176 watts, with lamps that cost $1 each. Upgrade B uses one 2 × 4, 4-lamp luminaire rated at 109 watts, with lamps that cost $3 each. Average energy cost is $0 .07/kWh. The annual hours of use for the system is 3500 hours. The burn cycle is 10 hours per day for 7 days each week. The facility is operated for 50 weeks each year. Lamp life is rated at 20,000 hours at 3 hours/start, but this value is extended to about 26,000 hours at 10 hours/start. Labor cost to replace lamps is $6 per lamp. (The value "3 hours/start" represents the average time lamps are on each time they are started before they are turned off—usually for 20 minutes—before being turned back on. See also "Burn cycle" in the glossary and Fact Sheet 12 in Chapter 8.)

$$\text{LOC} = \text{IC} + \text{LSL}(\text{AEC} + \text{AMC} = \text{ALC}) \tag{3.4}$$

where:

$\text{LOC}$ = life operating cost

$\text{IC}$ = initial cost

$\text{LSL}$ = lighting system life

$\text{AEC}$ = annual energy

$\text{AMC}$ = annual material cost

$\text{ALC}$ = annual labor cost

System A:

$$\text{AEC} = 176\ \text{W}/1000\ \text{kW} \times 3500\ \text{hours} \times \$0.07/\text{kWh} = \$43.12$$

$$\begin{aligned}\text{AMC} &= \text{lamps/year} \times \text{cost/lamp}\\ &= 4\ \text{lamps} \times (3500\ \text{hours/year} \div 26,000) \times \$1/\text{lamp}\\ &= 0.54\ \text{lamps/year} \times \$1/\text{lamp} = \$0.54\end{aligned}$$

$$\begin{aligned}\text{ALC} &= \text{lamps/year} \times \text{cost/lamp}\\ &= 0.54\ \text{lamps/year} \times \$6/\text{lamp} = \$3.24\end{aligned}$$

$$\text{LSL} = 10\ \text{years}$$

$$\begin{aligned}\text{LOC}_\text{A} &= \$40 + 10(\$43.12 + \$0.54 + \$3.24)\\ &= \$509.00\end{aligned}$$

System B:

$$\text{AEC} = 109\ \text{W} / 1000\ \text{kW} \times 3500\ \text{hours} \times \$0.07 / \text{kWh} = \$26.70$$

$$\text{AMC} = 0.54\ \text{lamps / year} \times \$3 / \text{lamp} = \$1.62$$

$$\text{ALC} = 0.54\ \text{lamps / year} \times \$6 / \text{lamp} = \$3.24$$

$$\text{LSL} = 10\ \text{years}$$

$$\begin{aligned}\text{LOC}_\text{B} &= \$80 + 10(\$26.70 + \$1.62 + \$3.24)\\ &= \$395.60\end{aligned}$$

The two LOCs may be compared by subtraction:

$$\begin{aligned}\text{LOC}_\text{A} - \text{LOC}_\text{B} &= \$509.00 - \$395.60\\ &= \$113.40\end{aligned}$$

Over its 10-year life, System A, the existing luminaire, costs $113.40 more than Upgrade B.

## Effects of Lighting Systems on HVAC Systems

The effects of the existing lighting systems on the building HVAC systems should be evaluated at this level. The installation of more efficient lighting systems will reduce the air-conditioning load. This reduction will decrease the electric bill for buildings using electric chillers. In some cases, this reduction can eliminate or forestall the installation of additional chiller capacity. This improvement can provide considerable savings for the building owner and is especially effective in downsizing some chiller plants that are being replaced as part of a change from CFC refrigerants.

The effect on the building heating systems is to increase the heating requirement. The analysis should ensure that there is sufficient capacity in the building heating equipment. The additional fuel cost should be calculated and subtracted from the cooling system savings.

In order to determine lighting-related HVAC energy, it is important to answer the following questions:

How much heat is generated by the lighting system?
Where does the heat go?
How does the heat affect the energy consumption of the HVAC system?

Lighting-related HVAC energy is calculated as follows:

$$\begin{aligned}\text{Lighting}-\text{related HVAC energy (kWh)} &= \\ \text{direct lighting energy (kWh)} &\times \\ \text{\% of year HVAC system operates} &\times \\ \text{\% of lighting heat impacting HVAC load} &\div \\ \text{\% efficiency of HVAC system}\end{aligned} \tag{3.5}$$

Not all of the heat from the lighting systems will have an impact on the HVAC load. The actual percentage that affects the HVAC load depends on the specific type of HVAC system and can be estimated with the help of an HVAC engineer.

Cooling efficiency may be calculated using the following equations:

$$\text{Efficiency} = \frac{\text{energy units delivered to or removed from space}}{\text{energy units into system}} \tag{3.6}$$

$$\text{Cooling efficiency} = \frac{\text{SEER (BTUh/W} - \text{h)}}{3.413\ \text{BTUh/W} - \text{h}} \tag{3.7}$$

SEER is the "seasonal energy efficiency ratio," which means the total cooling output of an air conditioning system during its normal usage period for cooling, in BTU/h, divided by the total electric energy input during the same period, in watt-hours. In the denominator of Equation 3.7, 3.413 is a conversion factor; there are 3.413 BTU per watt (BTUh/W-h). Efficiency therefore becomes unitless and is usually expressed as a percentage.

Research funded by the Electric Power Research Institute and reported in several technical publications, including EPRI Lighting Bulletin No. 6, April 1994 (MI-103646), shows how to calculate lighting and HVAC interactions for commercial offices. This lighting bulletin can be obtained from EPRI member utilities, and the information in the bulletin may also be found in the September 1996 issue of the IESNA magazine *LD+A*. Included in the publication is a table of U.S. locations and the "Fractions of Annual Lighting Energy to Cooling and Heating" that shows the portion of the year HVAC systems are estimated to operate and the fraction of lighting energy that goes to cooling or heating. For example, for Atlanta, Georgia, the additional air-conditioning benefit is listed at 19%. Another source for this data is an article in *Strategic Planning for Energy and the Environment* entitled "The Domino Effect."

## The Effect of Income Tax on Savings

The effect of income taxes is to reduce gross savings. The calculation is shown here:

$$\text{Net savings} = \text{gross savings} - \text{tax} \tag{3.8}$$

### Example of After-Tax Savings without Depreciation

$$\text{Gross savings} = \$100$$

$$\text{Tax rate} = 40\%$$

$$\text{Increment in taxable income} = \$100$$

$$\text{Tax} = \text{increment in taxable income} \times \text{tax rate}$$

$$= \$100 \times 40\% = \$40$$

$$\text{Net savings} = \text{gross savings} - \text{tax}$$

$$= \$100 \times \$40 = \$60$$

Depreciation is an expense that reduces income and lowers taxes.

$$\text{Increment in taxable income} = \text{gross savings} - \text{depreciation} \tag{3.9}$$

### Example of After-Tax Savings with Depreciation

$$\text{Gross savings} = \$100$$

$$\text{Tax rate} = 40\%$$

$$\text{Depreciation} = \$50\ \text{(for the current year)}$$

$$\text{Increment in taxable income} = \text{gross savings} - \text{depreciation}$$

$$\text{Increment in taxable income} = \$100 - \$50 = \$50$$

$$\text{Tax} = \text{increment in taxable income} \times \text{tax rate} = \$50 \times 40\% = \$20$$

$$\text{Net (after} - \text{tax) savings} = \text{gross savings} - \text{tax}$$

$$= \$100 - \$20 = \$80$$

## The Report

A comprehensive report should describe the condition of existing lighting systems and several upgrade options, including relighting, with a detailed analysis of each. Both positive and negative aspects of each measure considered should be reported so that the client can make an informed choice. The recommendations made in the report should also match comprehensive solutions to current task needs and provide flexibility for future needs. The maintenance savings calculated are reported, as well as the potential for productivity improvements. A detailed evaluation of lighting quality improvements should be outlined, including space improvements, such as refinishing surfaces. A description of the new appearance of the upgraded or relighted spaces may be included, supported by drawings or advanced graphic computer renderings. (Sample audit reports are provided and discussed in detail in Chapter 7.)

Sometimes remodeling plans coincide with lighting upgrades. When auditors discover these situations, they need to educate the building owner/manager of the effect that their choices of materials and colors will have on the light level and the luminance in the space. Renewing deteriorated room surfaces will increase light levels and provide a more attractive work environment. Choosing light colored surfaces and balancing the lighting are proven ways to provide effective lighting improvements. The perceptive analyst will provide calculations showing the effect of renewing the deteriorated room surfaces and the effect on light levels and the resulting energy savings when fewer luminaires are required, for example, for a relighting project.

## Summary

The purpose, reason, or need for performing a survey or audit will help determine which type is required for each project. The intent of the walk-thru survey, the simplest type, is to collect just enough information in a short period of time to make effective recommendations. The lowest-level walk-thru survey is performed to gather quickly the most meaningful facts needed for a one-to-one retrofit. For the most basic walk-thru survey, no fixtures are counted and no calculation of power density is made. The surveyor briskly tours the building in search of simple energy conservation measures (ECMs). Walk-thru surveys often indicate the need for a more extensive audit. Table 3.3 identifies what data is required for the walk-thru survey and how it is collected.

An intermediate audit is usually a hybrid of a walk-thru survey and a comprehensive audit, containing elements of each. An intermediate survey is more detailed than a walk-thru survey, and the data collection is usually done by a small team of two or three auditors. Facilities with numerous fixtures of the same type, laid out in repeated patterns, are usually audited using the intermediate audit. Auditors record their subjective judgments of the visual tasks performed, the state of maintenance routines, the surface finished, and the overall lighting quality. The data required in an intermediate audit and how it is collected is outlined in Table 3.3.

**TABLE 3.3** How Data is Obtained for Each Type of Survey or Audit[a]

| Data to Collect | Walk-Thru Survey | Intermediate (Mini-) Audit | Comprehensive Audit |
|---|---|---|---|
| Room dimensions | E or D | D or M | M |
| Equipment inventory | E or C | E or C | C |
| Lighting load | E | E or M | M |
| Hours of use | E or I | I or M | M |
| Lighting system voltage | I or D | D or M | M |
| Light level (FC) | O | E or M | M |
| Switching methods and zoning | O | D or C | C |
| Lighting problems | I | I | |
| Visual tasks | J | J | |
| State of maintenance | J | J | |
| Surface conditions | J | J | |
| Lighting quality | J | J | |
| Power quality | O | | |
| Daylighing potential | O | | |

[a]*C = count; D = (obtain from) drawing; E = estimate; I = interview; J = judgement (subjective); M = measure; O = optional*

# Chapter 4
# Tools and Measuring Instruments

Auditors can use many tools and measuring instruments to help them perform audits and surveys. Several tools are necessary, while others are optional investments that may be desired to shorten the time and reduce the expense of conducting an audit. Selecting the right tools and metering equipment is important because such equipment leads to increased accuracy, shortened audit times, and lower audit cost.

## Personal Safety Equipment

Safety equipment should be a primary consideration of the audit team. The objective is to protect the auditors while on the client's property, so *safety regulations of the client company must be followed.* Auditors should familiarize themselves with the common environmental hazards and contaminants found in each facility. It is not unusual for a client to require that the audit team watch a safety video or listen to a safety engineer explain where hazardous areas are located and what safety equipment is required in each area. Appropriate protective clothing including safety shoes is essential. A hard hat, safety glasses, and hearing protection may be required in some industrial environments (Figure 4.1). A respirator may be required in hazardous areas.

Be careful when working around rotating equipment. Never work alone and never work around equipment you are not familiar with, whatever your safety equipment. The client should have a policy about opening electrical power panels. If there is no written policy, ask for one and express your concerns. Many companies do not want anyone opening power panels due to the safety risk and the risk of power interruptions. Testing of electrical equipment should be performed by trained and experienced electricians who are thoroughly knowledgeable about the lighting equipment and systems. Voltages and currents are present that may cause serious injury or extensive damage. If you can enlist a company electrician to help, this arrangement will probably work best for everyone.

## Access Equipment

Where access to lighting equipment at low mounting heights is needed, a 6- or 8-foot step ladder is useful. For electrical safety reasons, use ladders made of fiberglass or other nonconductive materials. To access industrial or warehouse lighting, a scissors lift (for heights up to about 22 feet) or a highlift can be used. Note that the client company may not allow you to use its equipment due to liability concerns. Plan ahead and ask client representatives what their company policy is regarding use of their equipment. An alternative to using the client's access equipment is to arrange to have a company electrician use the company's ladders and lifts to help with the audit.

Common hand tools such as screwdrivers will be needed to open lighting equipment and to remove box covers. A bulb remover with a lightweight telescoping handle is handy for removing incandescent A, R, and PAR lamps from downlights.

## Data Recording Instruments

For data collection, a well-prepared set of preprinted audit forms will eliminate the need to return to the facility later to get information that was overlooked. Chapter 10 provides several forms for many different types of surveys and audits. Although many experienced auditors prefer to work with a blank notepad and collect only data pertinent to their analysis and recommendations, using forms will prevent missing data. A tape recorder works well when working alone

**FIGURE 4.1** Hard hat, safety glasses, and hearing protection, which may be required in some industiral environments.

or in small teams and where forms are not used. Photos and videos are both useful in complex facilities for reminding the audit team of what they saw during the site visit and for recreating some peculiar situations. Many companies require visitors to obtain advance approval before taking photographs or videos.

## Measuring Instruments and Their Use

### Measuring Distance

Data describing the space and its dimensions are required for several reasons. One is to calculate power density, in watts per square foot. Another is to verify light levels from zonal-cavity calculations. Space dimensions that are readily available from up-to-date drawings should be used; note, however, that it is not unusual to find that facility drawings are not current, and thus the information they contain may not be reliable.

In spaces with suspended, modular ceilings, counting ceiling tiles is a quick way to determine room sizes. Simply count the tiles in the lengthwise direction and multiply by the length of the tile. Then, count the tiles in the widthwise direction and multiply by the width of the tile. Sometimes there are nonstandard tile sizes, such as 20 inches wide or a special length such as 5 feet, so it is a good idea to measure the first tile before starting the count.

A fiberglass measuring tape (Figure 4.2) can be used to measure room dimensions, but this tool requires two people to use. A measuring wheel (Figures 4.3 and 4.4) makes the measurement of distance easier, but accurate use requires some open space in the room and at the walls to get the wheel up to the wall. Walking a straight line is an auditor skill that can prove helpful in obtaining accurate results. Care should be taken in reading the digital counter. Some wheels use scales that read in feet and inches, while others read in feet and tenths of feet. Measuring fixture spacings located in plaster ceilings or mounted in manufacturing spaces can be expedited by using a measuring wheel at floor level while eyeballing fixture centers.

**FIGURE 4.2** Large fiberglass measuring tape.

**FIGURE 4.3** A measuring wheel is easier to use than a measuring tape, which requires two people to obtain accurate measurements.

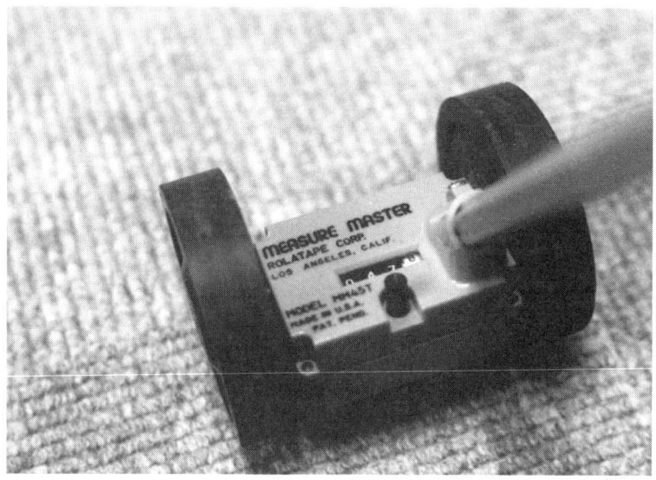


**FIGURE 4.4** Digital counters on measuring wheels must be read carefully since some use scales based on feet and inches, while others record feet and tenths of feet.

Using portable ultrasonic (sonar) or laser ranging tools is the fastest way to measure distance. Labor costs are reduced because only one person is needed to make the measurements and the user does not spend a lot of time walking the space. These electronic distance measuring tools make the measuring task up to 10 times faster than using measuring tapes or wheels in offices filled with furniture or in industrial spaces filled with machinery.

The ultrasonic ranging tools are small, portable, and rugged. They send narrow beams of high-frequency sound waves in the 25–40 kHz range that bounce off solid objects back to the handheld receiver. These tools measure the time from transmis-

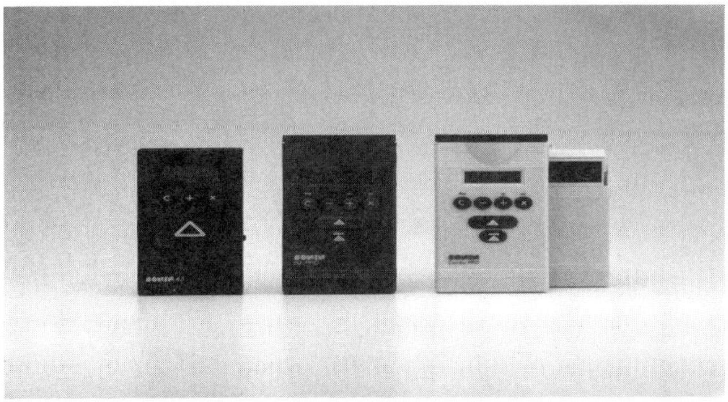

**FIGURE 4.5** Three comercial ultrasonic measuring tools available from SONIN.

sion of the sound waves to their reception at the handheld receiver and calculate distance on the basis of elapsed time. The distance calculated is displayed on a digital display. Ultrasonic ranging tools can measure, add, subtract, and multiply and can compute the square footage of entire buildings at the press of a button.

There are inexpensive ultrasonic units available that are designed for home use. These units are sold under the brand names ESTIMATOR, PROMEASURE, and HOME CONTRACTOR. They are not appropriate for most commercial audit applications because they have a limitation of about 30 feet.

Three commercial units are available from SONIN that are appropriate for lighting audits and cost approximately $150 to $250. The model numbers identify the maximum measurement. Thus, the SONIN 45 has a maximum measurement of 45 feet, and the SONIN 60 PRO measures a maximum of 60 feet. The SONIN Combo PRO when used alone will measure from 1.5 feet to 60 feet. A second electronic target unit can be added to extend the measuring distance to 250 feet in the dual unit mode. These tools (shown in Figure 4.5) measure in English units (feet and inches, decimal feet, or yards) or metric units (meters and centimeters). Linear measurements may be added or subtracted, and there are two memories for storing important measurements. Length and widths can be multiplied to obtain area.

Both the ultrasonic and the laser ranging tools can also be used to accurately measure vertical distances such as ceiling heights, building trusses, and industrial fixture mounting heights. The laser tools are faster and easier to use and yield better results than the ultrasonic tools when measuring to difficult shapes, such as cylindrical or tilted surfaces.

The Leica laser ranging tool (Figure 4.6) is small, rugged, and lightweight and uses an eyesafe Class 2 laser diode, which makes taking measurements with the unit safe. This tool has a range from 8 inches to normal targets at 100 feet, with accuracy to within +/- 1/8 inch. Using a small target board included with the instrument, the range can be extended to 330 feet. The target board can also help with measurements taken in bright light. The unit can measure from its front edge or its rear edge, and with the tracking function, measurements can be made continuously. Measurement time is approximately 3–10 seconds. An advantage of the laser unit is that you can see the point of measurement where the red laser spot hits the surface. Laser spot diameter is 0.25 inches at 33 feet and 1.2 inches at 165 feet. The display reads in either feet or meters and is illuminated for easy viewing.

Laser units are more expensive than ultrasonic units. For example, the list price of a Leica DISTO is $1395. A DATA DISTO RS232 model enables survey data to be transferred to and from PC's, laptops, and penpads. This model enables survey data to be recorded and post-processed more efficiently and less expensively than can be done using any of the manually read units.

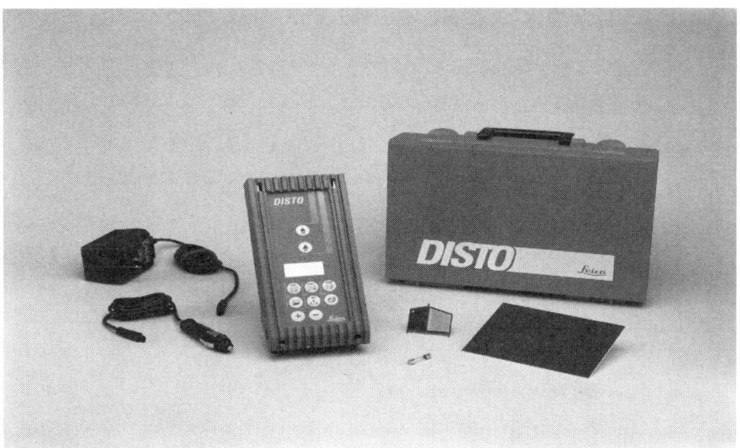


**FIGURE 4.6** Leica DISTO laser ranging tool.

## Counting Fixtures

In many cases, fixtures are installed in regular patterns of rows and columns and thus counting fixtures simply involves counting the number of rows and the number of columns and multiplying to get the total number of fixtures. Inexpensive hand-activated resettable tally counters, such as the one shown in Figure 4.7, are helpful, especially when counting large numbers of fixtures. Some auditors carry two counters, one in each hand, to count two different fixture types as they survey the area. It is important to remember to reset the counters after each area is counted; otherwise, an erroneously high number of fixtures will result.

Some fixture layouts are arranged in repeating bays. When this arrangement is encountered, count the number of fixtures in each bay and then multiply by the number of bays. Watch for the end of lineups, such as at walls, where partial bays may be encountered.

One audit firm uses commercial bar code equipment to manage the task of identifying and counting fixtures. An initial walk-thru is used to identify each fixture type, and then each type is assigned a unique bar code. Each auditor carries a sheet that gives a description of each type of fixture with its assigned bar code next to it. When counting fixtures, the auditor uses a handheld scanner and simply scans the bar code associated with each fixture encountered. The count is thus automatically tabulated by the bar code scanning equipment.

## Measuring Illuminance - Light Meters

The quantity of light or light levels are measured in English units of illuminance called footcandles or, in SI units, lux. One footcandle is equal to one lumen distributed evenly over a one-square-foot surface. One lux is equal to one lumen over a one-square-meter surface. Illuminance, in footcandles or lux, is measured with light meters, also called illuminance meters.

Analog meters consist of a light-sensitive barrier-layer cell and either a pivot and jewel or an analog taut-band meter movement. Figure 4.8 shows a typical analog illuminance meter. These meters are inexpensive and convenient to use, but they are not designed to be precision instruments: the accuracy is +/- 15% of full scale for the taut band and +/- 20% of full scale for the pivot and jewel movements. They are not rugged and thus need to be treated carefully and calibrated frequently to maintain reliability.

Digital illuminance meters, such as those shown in Figures 4.9 and 4.10, consist of either a selenium photocell or the more contemporary silicon photodiode cells and electronic circuitry (some contain a microprocessor). These compact battery-powered meters give the light level in footcandles (or lux) on a digital liquid crystal display. Some models are auto-ranging, while others have a hold feature, which is handy when measuring areas where the user cannot directly observe the display. Most are both cosine-corrected and color-corrected. Some models have a memory to hold multiple readings and come with data acquisition software and a serial port cable for downloading into a PC.

Light striking the light-sensitive cell face obliquely (from a slanting or inclined direction) produces illumination proportional to the cosine of the angle of incidence, but it may not produce an equivalent response from the cell. Since the meters are calibrated with light normal (at right angles) to the surface, light from oblique angles will yield lower readings than the true value unless correction is applied. Errors may vary from a few percent to 25%. A cosine-corrected meter has very nearly true cosine response at all angles of incidence.

The spectral response of the silicon cell is designed to closely approximate the CIE (Commission

**FIGURE 4.7** Hand-activated resettable tally counter.

**FIGURE 4.8** Analog illuminance meter.

Internationale de l'Éclairage) relative photopic luminosity curve so that the illuminance meter will evaluate luminous flux in the same way as the eye does. An uncorrected cell has a greater spectral sensitivity to wavelengths both in the violet-blue and red-orange regions than does the eye. Also, the cell responds to UV radiation and the eye does not. Meter manufacturers integrate a color filter with the cell so that the spectral sensitivity more closely approximates the CIE relative photopic luminosity curve to within 2%.

**FIGURE 4.9** Digital illuminance meter.

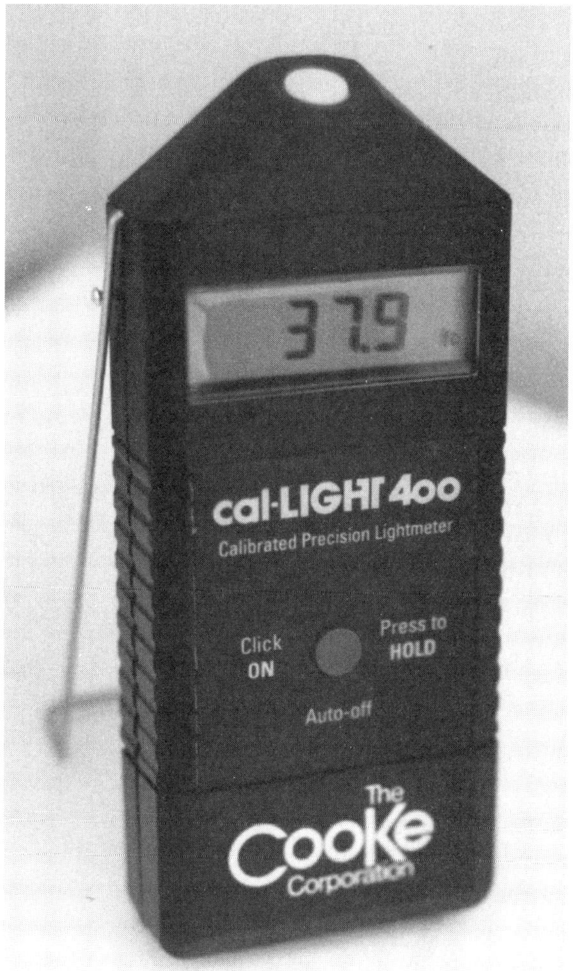


**FIGURE 4.10** Digital illuminance meter.

Because the response of light-sensitive cells to the various wavelengths of the visible spectrum is quite different from that of the human eye, meters without color correction read accurately only for the kind of illumination with which they were calibrated, usually using a filament lamp at a color temperature of 2700K. Most illuminance meters are made with a color-correcting filter that changes the cell response to a spectral distribution with sufficient accuracy for audit purposes.

## Light Meter Suppliers

AEMC Instruments
99 Chauncy Street
Boston, MA 02111
1-800-343-1391

The Cooke Corporation
600 Main St.
P.O. Box 888
Tonawanda, NY 14150-0888
Phone: 716-833-8274
FAX: 716-836-2927

EXTECH Instruments
335 Bar Hill Road
Waltham, MA 02154-1020
Phone: 617-890-7440
FAX: 617-890-7864

GE Lighting
Nela Park
Component #4162
Cleveland, OH 44112
1-800-626-2000, selection 3

Graseby Optronics
12151 Research Parkway
Orlando, FL 32826
Phone: 407-282-1408
FAX: 407-273-9046

Minolta Corporation
101 Williams Drive
Ramsey, NJ 07446-1293
Phone: 201-825-4000
FAX: 201-423-0590

OSRAM SYLVANIA, Inc.
Product Literature Group
c/o The Hibbert Group
400 Pennington Ave.
Trenton, NJ 08650
Phone: 1-800-453-1506
FAX: 1-800-453-1507

Tektronix, Inc.
P.O. Box 500
Beaverton, OR 97077
1-800-426-2200, option 1

Topcon Instrument Corp of America
65 West Century Road
Paramus, NJ 07652
Phone: 201-261-9450
FAX: 201-387-2710

## Measuring Illuminance - Techniques

Since many visual tasks are performed in a horizontal plane, illuminance measurements are made with the light meter in the horizontal position at a desk or bench top. Take readings where tasks are performed with workers in their normal work positions. However, the worker should be away from an area when the average light level in a space is being measured. Take several readings and calculate an average for comparison to values recommended by the Illuminating Engineering Society of North America (IESNA). The IESNA handbook outlines exact grid locations at which to set the meter to precisely measure the average illuminance of rooms. Note that while multiple measurements taken on a grid pattern are highly accurate, they are too time-consuming for most lighting audits.

Corridors are measured at the floor level. To obtain the average light level in a corridor, measure directly under one luminaire at the floor level. Take a second reading between luminaires. Calculate the average light level as shown in Fact Sheet 12 in Chapter 8.

Shadows and reflections caused by the position of the auditor with respect to the light meter must be eliminated. Some illuminance meters are designed with a separate cell paddle and are connected to the display unit by a cord. This type of meter allows the auditor to read the meter without casting a shadow on the cell. One hour should be allowed for fluorescent and high-intensity discharge (HID) systems to stabilize before measurements are made. When fluorescent or HID lighting systems with new lamps are measured, the lamps should be burned in for 100 hours of continuous operation before measurements are made. New incandescent sources require about 20 hours of burn-in.

An advantage of making measurements when the building is occupied is that there is no need to light up the building after normal working hours in order to find waste! If the audit is done after hours, someone should go ahead of the audit team to ensure that the lights are on at least an hour before illuminance measurements are made. Factors that can influence the illuminance measurements, such as lamp type and depreciation, surface reflectances, and lighting system voltage, should be recorded when it is desirable to match field measurements with zonal-cavity calculations. Illuminance measurements taken at the visual tasks will provide important data. These measurements are most easily done during occupancy, since it is easier to ascertain the visual tasks performed in a space when those who perform them are there.

Special techniques are required to account for the presence of daylight. First, measure the illuminance in several locations with the electric lights on. Next, turn off the electric lights and repeat the measurements in the same locations. This second set of readings is the daylight contribution. Subtract the two readings for each location to obtain the contribution from the electric lights. Note that if the electric lights are turned off first to measure the daylight contribution, and then turned on to get the combined contributions of the electric lights and the daylight, fluorescent lamps (and HID lighting) will usually not be stabilized enough for measurements to be meaningful.

To determine if fluorescent fixtures contain a magnetic or an electronic ballast, a simple flicker checker can be used. This is a variation on a children's spinning top that has markings that appear to strobe when observed under fluorescent lighting with magnetic ballasts. Under fluorescent lighting equipped with high-frequency electronic ballasts, the markings appear stationary. Figure 4.11 shows a flicker checker available from Motorola Lighting (1-800-MLI-0089).

## Measuring Electrical Values

### Lighting System Voltage

It is important to obtain lighting system voltage values so that the correct ballasts can be ordered for the upgrade project. In addition, several spot checks of the building lighting system voltages will determine if the applied line voltage is within tolerance. The effect of over and under lighting system voltage is detailed in Fact Sheet 10 in Chapter 11. Although building receptacle and lighting circuits are usually supplied from different transformers, a spot check of receptacle circuits can tell if out-of-tolerance building voltages are present.

A simple, inexpensive (about $30) plug-in expanded-scale voltmeter, shown in Figure 4.12, is a handy tool that can be used as a screening tool. This meter has a range of 100 to 130 VAC (volts alternating current) and an accuracy of +/- 3%. If the plug-in meter shows a reading that is higher or lower than normal, then lighting system voltages should be measured at the lighting distribution breaker panels. A digital multimeter makes the measurement of lighting system voltages easy.

Some building owners will not allow an audit team to make measurements at the lighting panels. Determine the specific building policy before starting the on-site audit. In some cases, building owners will allow the panels to be opened by a qualified, authorized electrician. There may also be restrictions on opening panels during normal working hours because of the liability of inadvertent work disruptions and safety concerns.

### Lighting Loads

Measuring lighting loads is preferred to calculating or estimating lighting loads because (1) there is usually a mix of ballast vintages and lamp types and (2) ANSI test data and actual field measurements of power can differ by as much as 25%. It is not unusual to find both 40-watt and 34-watt lamps and two or three ballast vintages in the same facility. The

**FIGURE 4.11** Simple flicker checker, used to determine if a fluorescent fixture contains a magnetic or an electronic ballast.

**FIGURE 4.12** Plug-in expanded-scale voltmeter.

exact mix cannot be calculated unless each fixture is opened to determine the lamp and ballast combination for each fixture. It is more cost-effective to measure the lighting loads than opening fixtures.

The lighting system current is sometimes measured using a true r.m.s. (root-mean-square) clamp-on ammeter, calibrated in watts, instead of using a power meter. Current and power will maintain a close relationship for most applications, providing accurate data. The r.m.s. clamp-on is recommended because average-sensing meters can produce readings that are 30–50% low when harmonics are present. Using the amp reading from the clamp-on meter and multiplying it by the applied voltage is not acceptable. The reasons that the simple volt-amp (VA) product is not acceptable are because of the power factor and the applied voltage. Most building lighting voltages vary from the nominal 120 or 277 voltage, so when the nominal voltage is used, the product is incorrect (see Chapter 8, Lighting Fact Sheet 10).

## Clamp-on Ammeter Manufacturers

AEMC Instruments
99 Chauncy Street
Boston, MA 02111
1-800-343-1391

Amprobe Instrument
630 Merrick Rd.
Lynbrook, NY 11563
Phone: 516-593-5600
FAX: 516-593-5862

B&K Precision
6470 West Courtland
Chicago, IL 60707
Phone: 312-889-1448
FAX: 312-794-9740

EXTECH Instruments
335 Bar Hill Road
Waltham, MA 02154-1020
Phone: 617-890-7440
FAX: 617-890-7864

Fluke Corporation
P.O. Box 9090
Everett, WA 98206
1-800-44-FLUKE

LEM Instruments, Inc.
1711 Elmhurst Rd.
Elk Grove Village, IL 60007
1-800-536 6775

UEI
5500 Southwest Arctic Dr.
Beaverton, OR 97005
Phone: 1-800-547-5740
FAX: 503-643-6322

Yokogawa Corporation of America
2 Dart Road
Newnan, GA 30265
Phone: 404-253-7000
FAX: 404-251-2088

## Power Meters

A true r.m.s. clamp-on wattmeter that measures both the applied voltage and the lighting load current and that takes into account power factor will be useful for more detailed measurements. A power/ demand analyzer can provide single- or multiphase, single circuit or whole building data on electrical energy and demand. A power quality analyzer can add analysis of electric transients and harmonic distortion on the lighting circuits.

The wattmeter chosen must be compatible with the lighting systems being measured. The instrument must be rated to accurately read the range of applied voltages and load currents and the harmonics encountered. Be especially careful when measuring just one or two fixtures. Many power meters will not read accurately at light (low) loads.

There are a number of manufacturers of wattmeters, including some that provide clamp-on, handheld models. Some models include the measurement of harmonic distortion.

## Power Meter Manufacturers

AEMC Instruments
99 Chauncy Street
Boston, MA 02111
1-800-343-1391

DRANETZ Technologies Inc.
1000 New Durham Rd.
Edison, NJ 08818-4019
Phone: 908-287-3680
FAX: 908-248-1834

EXTECH Instruments
335 Bar Hill Road
Waltham, MA 02154-1020
Phone: 617-890-7440
FAX: 617-890-7864

Fluke Corporation
P.O. Box 9090
Everett, WA 98206
1-800-44-FLUKE

Yokogawa of America, Inc.
2 Dart Road
Newnan, GA 30265
Phone: 404-253-7000
FAX: 404-251-2088

### Measuring Hours-of-Use

The time that lighting systems are on is referred to as hours-of-use. Accurate hours-of-use lighting data is the most difficult audit information to obtain. The actual time lighting systems are operated are usually quite different than estimates. When the hours-of-use data is inaccurate it causes an inaccurate computation of kilowatt-hours (kWh). When kWh data is unreliable, it can distort the economic payback of a lighting upgrade. Estimates based on assumptions, rules of thumb, or the occupants best guess about lighting schedules are unreliable. Requesting this information from building owners or tenants is usually useless. Their answers to the auditor's questions concerning operating hours are usually inaccurate. For this reason, it is important for an auditor to try to obtain hours-of-use information firsthand, by observation or by measuring. For example, if you want to know if lights are being left on at night, then visit the facility at night. A midnight drive by the facility will help uncover this information and help you know where to set up time-measuring instruments for more details.

Time-measuring instruments are more accurate than verbal information in determining time of operation. Elapsed-time meters are electrical timing motors that drive a digital counter that is calibrated in hours. These instruments are simple, inexpensive contact instruments that can be connected to lighting circuits or lighting contactors. Since they are usually connected to a "hot" lighting circuit, these connections should be made by a qualified electrician. The time of connection and the meter reading are recorded as the beginning of the audit period. The audit period should be at least a full week, but a month is preferable. At the end of the audit period, the ending reading is recorded and the beginning reading is subtracted to obtain the hours of use. The time between readings is extrapolated to calculate the annual operating hours.

### Lighting Loggers

Lighting loggers make it possible to obtain hours of lighting use easily and inexpensively. Loggers are noncontact instruments so there is no need to have an electrician make connections to electrical equipment. The value of loggers is their ability to provide accurate and reliable data for use in analyses of lighting upgrades in a timely and cost-effective way.

Loggers are small, rugged, battery-powered electronic units, about 1 × 2 × 4 inches. The units come with various capabilities and built-in functions. In their most basic form, loggers include a photocell and a digital display. When the logger "sees" light, it counts time. These units use a crystal-controlled time base, so they have accuracy to one minute per month. A magnetic strip allows mounting inside a fixture to safeguard the unit against theft, as shown in Figure 4.13. A sensitivity adjustment may be provided to tune out ambient light from other sources. Loggers have a means of resetting the time to clear

**FIGURE 4.13** Mounting a lighting logger inside a fixture to discourage theft.

the display before beginning the next observation period.

The simplest units are elapsed-time totalizers that add all the time intervals the lights are sensed on during the observation period. The total shown on the digital display is the accumulation of all the hours that the lights are on (Figure 4.14). The more sophisticated units log each on and off event as well as time.

To select the correct logger for a particular audit requires a knowledge of the data needed. For example, if a utility has time-of-day electric rates, a logger that can record time-of-day usage is needed (Figure 4.15). If the main interest is in kWh savings, then a basic elapsed-time totalizing unit would be selected, such as that shown in Figure 4.16.

If the auditor is looking for areas where occupancy sensors may be used effectively to save energy, special loggers like the Watt Stopper InteliTimer Pro Logger or the MyTech Energy Monitor EM-301 can be used. The EM-301 employs an external portable

**FIGURE 4.14** Simple elapsed-time totalizer.

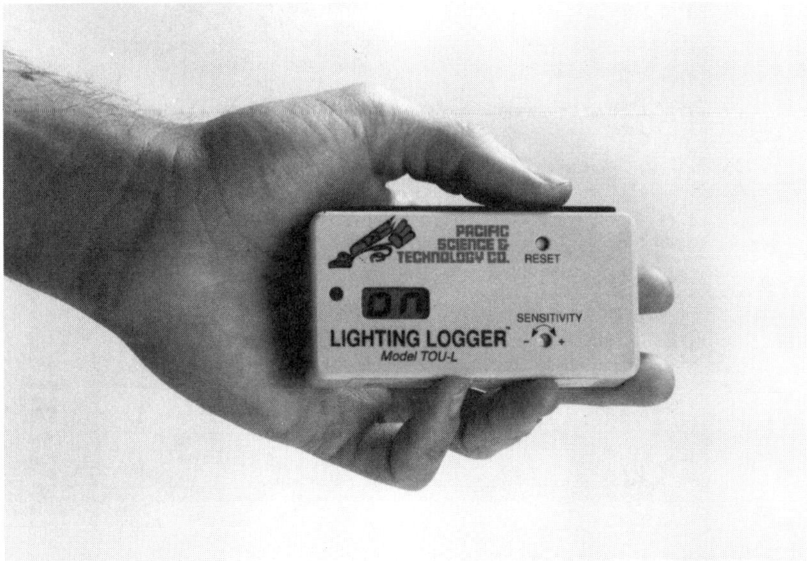


**FIGURE 4.15** Logger that records time-of-day usage.

occupancy sensor, while the Pro Logger has an integral PIR (passive-infrared) sensor. These loggers establish the energy-saving potential of using occupancy sensors by measuring the time a space is occupied and unoccupied and the time the lighting system is on and off during the audit period. The InteliTimer Pro logger (shown in Figure 4.17) uses PIR technology to detect occupancy of an area. This logger observes the light level through a clear plastic light pipe. The light level sensitivity is adjustable and a light level LED allows users to set the logger to identify the lighting on and off levels in different locations. This allows the logger to distinguish electric lighting from daylighting. Once in place, the Pro logger records a log entry every time there is a change in occupancy or lighting status and stores a detailed history for retrieval by a PC or Macintosh via a standard RS-232 serial port. The InteliTimer package includes a software package for either the PC Windows environment or Macintosh systems. The software analyzes and displays collected data through an analysis report as well as through graphic displays and printouts.

## Logger Data

Lighting loggers can record four types of lighting usage data:

1. **Runtime data:** One number that represents the total time that lights were on during the monitored period. This number is obtained using the simplest and least expensive logger, the totalizing logger (Figures 4.14 and 4.16), which adds all the time intervals during the monitored period.

**FIGURE 4.16** Elapsed-time totalizing logger unit.

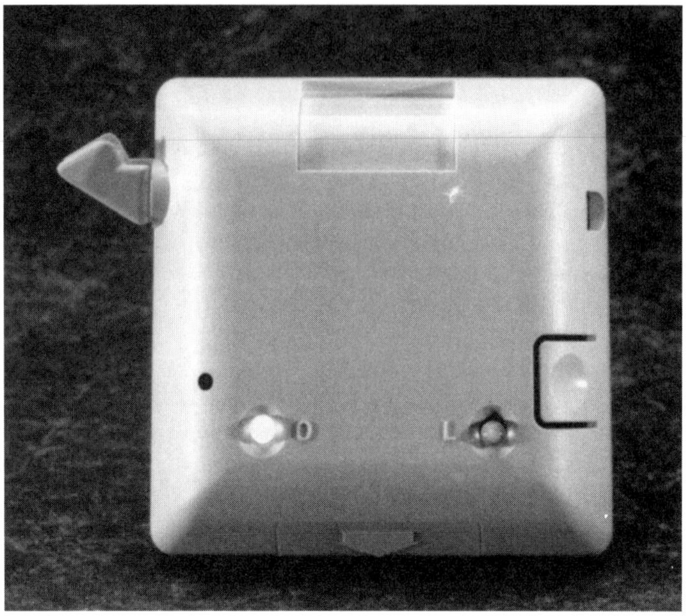

FIGURE 4.17

2. **Event counting data:** The number of times the luminaire being monitored was turned on.
3. **Time-of-use (TOU) data:** The times of day at which luminaires being monitored were turned on and off. TOU loggers (Figure 4.15) provide a complete usage pattern of a monitored luminaire or a room. Software, usually provided by the logger manufacturer, is required for processing the data collected.
4. **Occupancy data:** The number of hours a luminaire was on while a given room was unoccupied. Occupancy loggers are more sophisticated and more expensive than other logger types and are made in both run-time and time-of-use models.

### *Logger Features*

**Adjustments**—A sensitivity adjustment allows adjustment of the level of light required to trigger the lighting logger. Installing the logger in a luminaire makes it less sensitive to stray sources of light and prevents theft of the unit. A magnetic strip on the logger snaps the unit firmly in place.

**Fiber-Optic Cables**—Some loggers come equipped with a fiber-optic cable that restricts the logger's "field of vision" and helps eliminate triggering from stray light. The end of the fiber-optic cable can be directed to the bulb wall in order to better sense the light from a fluorescent bulb.

**Battery Life**—Battery life can range from 6 months to 10 years depending on the power draw of the logger. Loading fresh batteries before an audit will ensure that a battery failure does not interrupt data collection in the middle of a monitoring period.

### *Data Recording Methods*

There are three ways that loggers collect data. The simplest way is to read the digital liquid crystal display (LCD) and manually record the single reading. Totalizing runtime loggers are recorded in this way. The more sophisticated loggers are data recorders that track and store events and time intervals in an internal memory. These loggers require downloading of the stored data into a PC. At least one logger manufacturer provides additional channels to allow monitoring digital events or analog values to permit the correlation of lighting use with other variables or for other energy monitoring. One manufacturer uses an infrared scanner to download data from loggers. The scanner is then downloaded to a PC. This feature permits quick on-site data collection and does not require the removal and replacement of the logger.

### *Logger Software*

Software is provided by some logger manufacturers so a PC can be used to analyze and retrieve data from their loggers. A logger that has been on-site for an observation period can be brought to the PC and connected through the RS-232 serial port for direct PC access to the information. For remote access, the information can be downloaded via a modem over a telephone line. For direct PC access, the serial port is usually used for two-way communications with the logger. The logger's current status can be displayed, and the time and date can be set. Downloading the logger data is a simple matter of selecting a data-retrieve command. A file name is assigned based on the date and an identification number assigned by the auditor. After the data is downloaded, the logger memory can be reset through a PC command.

An analysis menu allows the user to view, analyze, and graph the data collected by the logger. Selecting a data file begins the analysis process. The software will present a list of files to choose from, or a file name may be typed in. The selected file will be the current file for the entire analysis session. The data can be exported to spreadsheet programs such as Lotus 1-2-3 or EXCEL.

The computer can be instructed to summarize the data to provide a statistical summary of data stored in the file, including the total and average on-times and off-times, minimums and maximums, and so on. A sample summary report is shown in Table 4.1. Transition data may be displayed in the order in which it was collected. A transitions report is shown in Table 4.2. Time series data may be displayed as a percentage of on-time for user-specified time intervals. This data is useful for locating peak load times. Table 4.3 shows a time series report.

The user can choose to analyze the data by rate schedule. This function is used to analyze usage patterns by collecting all of the run-time hours into user-specified time periods. For example, it may be desirable to associate on-peak, mid-peak, and off-peak periods. Run-times and other useful data are then calculated for each period. A summary report may be selected to view the summary statistics for each defined rate period. Each major period is summarized on separate screens.

Using the same software, computer graphs can be created to display data in a concise form that is easy to read and easy to analyze. For example, if the user selects to graph usage by rate schedule, each period can be shown as a different color, defined when the rate schedule was created. Another graph displays the collected on/off transition time data in a square wave, duty-cycle format. The date and time of each transition are displayed. The user may desire to display the collected data in 15-minute blocks of on-time for each day. Data are displayed in 15-minute increments.

**TABLE 4.1** Sample Logger Summary Report

| | |
|---|---|
| DATA FILE NAME: | C:\SMARTLOG\EXAMPLE.LOG |
| SERIAL NUM: | 930519-023 |
| DESCRIPTION: | Office Lighting, Suite 201 |
| Starting: | 03/08/1996 at 08:02 |
| Ending: | 04/05/1996 at 16:06 |
| Total Time: | 848.07 hrs |
| ON Periods: | 42 |
| Percent ON: | 38.4% |
| Total ON Time: | 325.43 hrs |
| Average ON Time: | 7.75 hrs |
| Longest ON Time: | 16.98 hrs |
| Shortest ON Time: | 1.43 hrs |
| OFF Periods: | 41 |
| Percent OFF: | 61.6% |
| Total OFF Time: | 522.63 |
| Average OFF Time: | 13.32 hrs |
| Longest OFF Time: | 40.95 hrs |
| Shortest OFF Time: | 3.03 hrs |

**TABLE 4.2** Sample Logger Transition Report

FILE: C:\SMARTLOG\EXAMPLE.LOG
SERIAL NUM: 930519-023
DESCRIPTION: Office Lighting, Suite 201
RECORDS: 84

| Record | Date | Time | Transition |
|---|---|---|---|
| 1 | 03/08/1996 | 08:02:00 | Turned ON |
| 2 | 03/08/1996 | 17:06:00 | Turned OFF |
| 3 | 03/09/1996 | 07:59:00 | Turned ON |
| 4 | 03/09/1996 | 17:00:00 | Turned OFF |
| 5 | 03/10/1996 | 20:02:00 | Turned ON |
| 6 | 03/10/1996 | 22:57:00 | Turned OFF |
| 7 | 03/11/1996 | 07:45:00 | Turned ON |
| 8 | 03/11/1996 | 18:30:00 | Turned OFF |
| 9 | 03/12/1996 | 00:14:00 | Turned ON |
| 10 | 03/12/1996 | 03:57:00 | Turned OFF |
| 11 | 03/13/1996 | 07:59:00 | Turned ON |
| 12 | 03/13/1996 | 17:03:00 | Turned OFF |
| 13 | 03/14/1996 | 08:10:00 | Turned ON |
| 14 | 03/14/1996 | 16:15:00 | Turned OFF |
| 15 | 03/15/1996 | 09:46:00 | Turned ON |
| 16 | 03/15/1996 | 18:05:00 | Turned OFF |
| 17 | 03/16/1996 | 12:02:00 | Turned ON |
| 18 | 03/16/1996 | 14:05:00 | Turned OFF |
| 19 | 03/17/1996 | 08:02:00 | Turned ON |
| 20 | 03/17/1996 | 17:06:00 | Turned OFF |
| 21 | 03/18/1996 | 07:59:00 | Turned ON |
| 22 | 03/18/1996 | 22:57:00 | Turned OFF |

**TABLE 4.3** Sample Time Series Report

FILE: C:\SMARTLOG\EXAMPLE.LOG
SERIAL NUM: 950519-023
DESCRIPTION: Office Lighting, Suite 201
RECORDS: 847

| Record | Date | Starting | Ending | Percent ON |
|---|---|---|---|---|
| 1 | 3/12/96 | 00:00:00 | 00:59:59 | 76.7 % |
| 2 | 3/12/96 | 01:00:00 | 01:59:59 | 100.0 % |
| 3 | 3/12/96 | 02:00:00 | 02:59:59 | 100.0 % |
| 4 | 3/12/96 | 03:00:00 | 03:59:59 | 95.0 % |
| 5 | 3/12/96 | 04:00:00 | 04:59:59 | 0.0 % |
| 6 | 3/12/96 | 05:00:00 | 05:59:59 | 0.0 % |
| 7 | 3/12/96 | 06:00:00 | 06:59:59 | 0.0 % |
| 8 | 3/12/96 | 07:00:00 | 07:59:59 | 1.7 % |
| 9 | 3/12/96 | 08:00:00 | 08:59:59 | 100.0 % |
| 10 | 3/12/96 | 09:00:00 | 09:59:59 | 100.0 % |
| 11 | 3/12/96 | 10:00:00 | 10:59:59 | 100.0 % |
| 12 | 3/12/96 | 11:00:00 | 11:59:59 | 100.0 % |
| 13 | 3/12/96 | 12:00:00 | 12:59:59 | 100.0 % |
| 14 | 3/12/96 | 13:00:00 | 13:59:59 | 100.0 % |
| 15 | 3/12/96 | 14:00:00 | 14:59:59 | 100.0 % |
| 16 | 3/12/96 | 15:00:00 | 15:59:59 | 100.0 % |
| 17 | 3/12/96 | 16:00:00 | 16:59:59 | 100.0 % |
| 18 | 3/12/96 | 17:00:00 | 17:59:59 | 5.0 % |
| 19 | 3/12/96 | 18:00:00 | 18:59:59 | 0.0 % |
| 20 | 3/12/96 | 19:00:00 | 19:59:59 | 0.0 % |
| 21 | 3/12/96 | 20:00:00 | 20:59:59 | 0.0 % |
| 22 | 3/12/96 | 21:00:00 | 21:59:59 | 0.0 % |

## Lighting Logger Manufacturers

Electronic Product Design
6962 Bluebelle Way
Springfield, OR 97478
Phone: 503-741-0778

MyTech Corporation
706 Brentwood Street
Austin, TX 78752
FAX: 512-450-1215
Phone: 512-450-1100

North Fork Retrofit (NFR)
511 Carpenter Street
Greenport, NY 11944
FAX: 516-477-2922

Pacific Science and Technology
64 NW Franklin Avenue
Bend, OR 97701
FAX: 503-385-9333
Phone: 502-388-4774

The Watt Stopper
2800 De LaCrus Boulevard
Santa Clara, CA 95050
FAX: 408-988-5373
Phone: 408-988-5331

XENERGY
Three Burlington Woods
Burlington, MA 01803-4543
FAX: 617-229-4867
Phone: 1-800-967-4393

## Locating Veiling Reflections

A special demonstration sheet, such as shown in Figure 4.18, may be used to verify the presence of veiling reflections, that are produced by lights mounted directly over a task and project light at a critical angle. Some of the light is reflected into the eyes of the viewer as veiling reflections, partially or totally obscuring the task details by reducing the contrast.

## Carrying Tools and Instruments

The auditor will need a way to carry all the instruments and tools described. Without some way of carrying all these items, an auditor will likely leave a tool or instrument in one of the spaces being audited. One way to organize and haul all the tools and instruments is by wearing a photographer's vest. The multiple pockets of various sizes provide a practical way to handle and keep track of all the equipment needed for the lighting audit or survey. Another way would be to carry a small toolbox or briefcase.

## Lighting Evaluation System (LES)

The LES is a computerized monitoring system consisting of portable data loggers and computer analysis software developed by Architectural Energy Corporation of Boulder, Colorado, under the sponsorship of the Electric Power Research Institute (EPRI) and six electric utilities. LES simplifies and automates lighting system performance evaluations and develops a description of the lighting system. LES develops a data collection plan and initializes a suite of small, battery-powered MicroDataLoggers (Figure 4.19), downloads the data, and automatically analyzes it. The system monitors dedicated lighting circuits and allows individual luminaire monitoring for circuits with mixed end uses. A complete monitoring solution is provided in one package. The four sensor inputs of each portable MicroDataLogger allow illuminance sensors, AC watt tranducers, and temperature sensors to be plugged into any channel. The analysis software develops lighting load shapes for both the pre-upgrade and post-upgrade scenarios, calculates differences (savings, including reduction in lighting power demand), predicts annual energy use, and describes the effect of the lighting on the HVAC system energy consumption. The LES may be used for lighting system commissioning and diagnostics as well as for lighting evaluations.

## Summary

Selecting the right tools and metering equipment is an important way to increase the accuracy of and decrease the time to perform surveys and audits, thus lowering costs. Protective clothing and personal safety equipment are important for the auditor's protection, especially in industrial locations. Access equipment such as lifts and ladders are needed to reach lighting systems for close inspection, and hand tools, like screwdrivers, are required for opening lighting panels. A bulb remover with a telescoping handle is handy for removing incandescent A, R, and PAR lamps from downlights.

When forms are not used, a tape recorder comes in handy for recording information as the auditor moves from space to space. Photographs and videos (if allowed by the client company) are useful in complex facilities for reminding audit team members what they saw on-site.

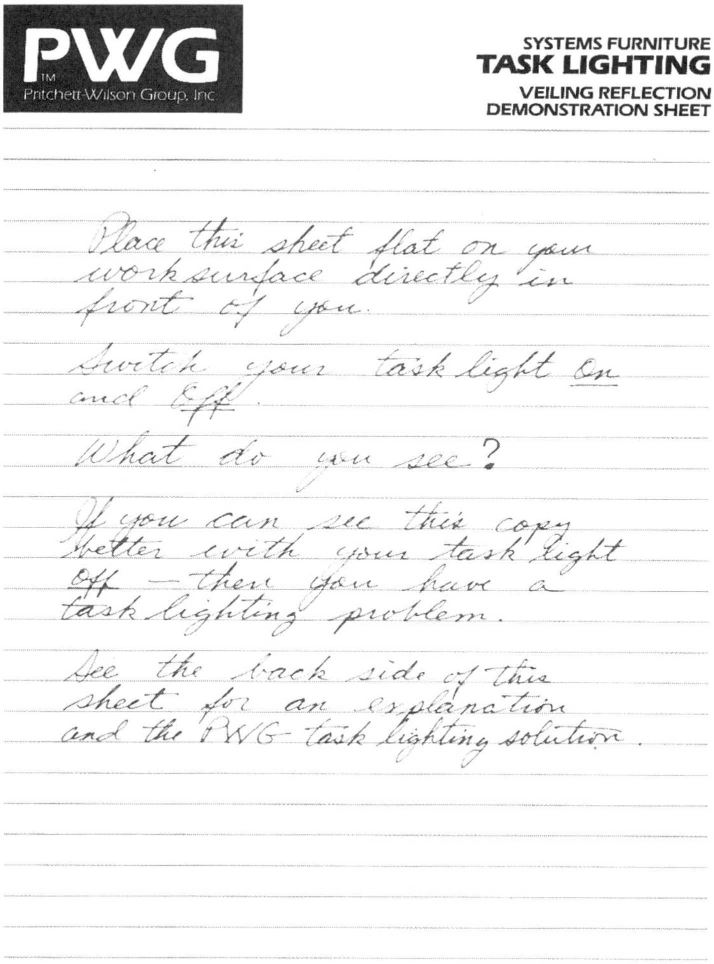

PWG
TM
Pritchett-Wilson Group, Inc.

SYSTEMS FURNITURE
TASK LIGHTING
VEILING REFLECTION
DEMONSTRATION SHEET

Place this sheet flat on your worksurface directly in front of you.

Switch your task light On and Off.

What do you see?

If you can see this copy better with your task light Off — then you have a task lighting problem.

See the back side of this sheet for an explanation and the PWG task lighting solution.

**FIGURE 4.18** This special demonstration sheet can be used to verify the presence of veiling reflections.

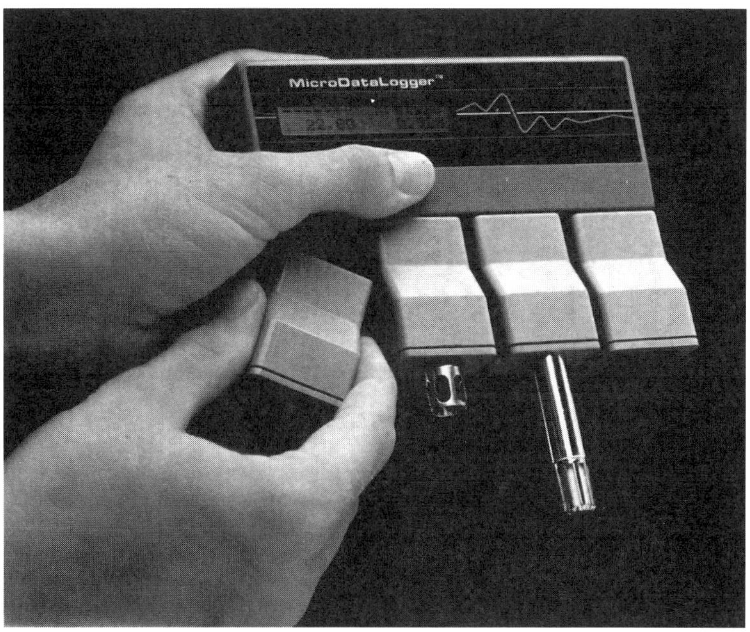


**FIGURE 4.19** Battery-powered MicroDataLogger.

A measuring tape or a measuring wheel may be used to measure room dimensions, but using ultrasonic (sonar) or laser ranging tools is the fastest way to measure distance. Hand-activated tally counters are helpful for counting large numbers of fixtures. Both analog and digital light meters are available to measure illuminance.

A simple, inexpensive plug-in expanded-scale voltmeter is a handy screening tool that can be used to determine if building voltage is within tolerance. Multimeters are used to measure lighting system voltages at lighting panels. Lighting loads may be measured with clamp-on ammeters or portable power meters. A computerized monitoring system called the Lighting Evaluation System (LES) consists of portable data loggers and computer analysis software. LES is used for large-scale comprehensive data collection.

Chapter 5

# Collecting Building Information: Tips and Techniques

The heart of any audit is data collection. This is usually regarded as the menial labor part of the audit process because of the drudgery involved. Data collection is hard work and requires endurance but it is the most vital part of an audit, because without reliable data, the other steps in the audit process will be flawed. Costly mistakes can be avoided by time and effort spent collecting relevant data at the building audit site.

## Scope of Data Collection

The goal of an on-site audit is to obtain solid, accurate information in a short time. The time spent collecting the right information in sufficient quantity (and no more) is the key to an effective audit. Enough time must be spent collecting the required data, but many auditors spend too much time. Spending more time to get more details is rarely cost-effective. Our firm does not use lighting professionals as auditors. Instead, our auditors are trained to focus on data collection. Professional lighting engineers and designers get their creative juices flowing too soon, interfering with the data collection process. Starting detailed plans for changes at this stage is premature and will take too much time. A lighting professional should lead the on-site audit and be there to answer auditors' questions, so the leader's focus should be on the big picture, not on data collection.

The amount of information required will depend on the type and purpose of the audit. The first chapter of this book outlined the three types of lighting audits—the walk-thru survey, the intermediate survey, and the comprehensive audit. Chapter 3 described in detail the information to be gathered in each type of audit. A walk-thru survey requires the least data, while a comprehensive audit requires the most data.

Remember, at the building site we are performing an audit, not conducting research studies or holding a design clinic. The idea is to know what to do and how to do it, before going on-site. The scope of data collection should be set before the on-site audit so on-site audit teams do not have to worry about making decisions concerning the amount of detail to collect and thus can focus on the task of data collection. For guidance, auditors can use the preprinted forms discussed in Chapter 3 and shown as sample forms in Chapter 10.

## Before the On-Site Audit

### Collecting Information

The first step in performing a successful lighting audit is to collect building information. While most of the building data will be revealed from the on-site data collection process, some building information—specifically, electrical data and building drawings—needs to be collected and reviewed before the audit team even goes to the building site. Client input in the form of details and preferences can help shape the audit plan and is helpful to have ahead of time.

#### Electrical Data

The correct electric rate schedule and electric bills for the building being audited should be obtained for the previous 12 months' billing records. If the client company does not have copies of the actual bills, obtain them from the electric utility. Correct meter numbers and a release letter from the client will help get these records. Client company records taken from the original bills will not reveal rate and meter ID information and should be used only if copies of the electric bills cannot be obtained.

Make sure that the bills being examined are for the correct meters. Matching the bills to the right meters at this stage will save much unrewarding analysis time later. Electric utilities provide many different rates for their customers. Checking the rate schedule that appears on the bills may raise questions about whether the client company is on the correct rate. This issue can be explored in some detail later, but at this point, a preliminary look may be beneficial to your client.

The applicable electric rate schedule is useful because components of the total cost of electric service are often difficult to identify from the monthly bills. If monthly peak demand charges are found on the bills, detailed records can be requested from the electric utility. These records will be used later in the analysis phase. Determining the way a proposed lighting change will affect the client company's electric bill is an important part of determining the true savings that will accrue from the project.

## Building Drawings

Building electrical drawings should be acquired to preview the building lighting systems and how they are arranged. These drawings can also be used to obtain the lighting system voltages (lighting systems in the same building may not have the same voltage). The electrical drawings will also help to identify if there are several different lighting fixture types or vintages of equipment. When this is the case, the team leader may direct the audit team to start the audit with a tour of spaces that contain representative fixtures and view all types before the detailed on-site audit begins. Drawings are also useful because knowing the lighting control locations before the on-site visit will reduce the time spent looking for controls in the dark!

Having architectural drawings can help reduce the number of area measurements required at the site. Drawings that are not up-to-date—sometimes called "as was" instead of "as built"—can easily be checked for accuracy/usefulness upon entering each space. A good review of all the applicable building drawings will help determine the scope of information to be collected on the preprinted forms.

## Client Input

Information such as a list of stockroom lamp (bulb) types, space remodeling plans, lighting upgrade plans, available rebates, and customer equipment preferences are requested from the client in an effort to limit the scope of information to be collected. The lamp (bulb) information can be helpful when sorting out bulb types during the audit. Be aware that there may be vintages of bulbs in the storeroom that are no longer in use in the building. This issue may need to be resolved with the client's representative before starting the on-site audit. Client preferences for lighting upgrade equipment are important to know. For example, if the client company is only considering one-for-one retrofits, then some information—such as ceiling heights and existing fixture spacings—will not be needed.

## Organizing the Audit

An organized plan identifying each auditor's task, prepared prior to the on-site visit, can reduce time spent on-site. The plan is written based on the age, size, and type of building being surveyed and on the type of lighting systems in the building. Manufacturing spaces in industrial buildings are not audited the same way as office spaces in high rises. Each requires a different organized plan of attack.

During this phase, the lighting professional or team leader must decide whether the on-site phase will be conducted while the building is occupied or unoccupied. This decision will most likely be influenced by the client's preferences, but our experience is that offices are best done during unoccupied periods and manufacturing areas are best done during occupied times. The reasons are based on the tasks being performed in the spaces. Office tasks are well known, while many manufacturing tasks are not, and a better understanding of the industrial tasks can be obtained while observing these tasks being carried out by the workers who normally do them. Also, most manufacturing facilities are less densely occupied than are office spaces.

On-site audits are done the same way as eating an elephant—one piece at a time. In the case of audits it helps to know which pieces to eat first. The plan of attack drawn up prior to the on-site audit outlines the procedures to be followed and the sequence. Access to certain spaces will help determine the sequence and schedule. A time schedule paces the audit teams and ensures completion on time.

We pay our auditors a bonus if the audit is completed sooner than planned and if it is up to our standards of accuracy. Accuracy is checked while the audit teams are still on-site and, if it seems to be substandard in particular areas, those areas are rechecked, usually by a different team. Since the auditors are paid an hourly rate, the bonus helps make up for lost wages due to early completion. The bonus thus keeps the audit from dragging on forever. Friendly competition between the audit teams can inspire their efforts and help reduce boredom.

Assigning the duties of each team member ahead of time will make expectations clear and help avoid confusion and mistakes. Building sketches or floor layouts are used to divide areas for each team and to set boundary rules. This is necessary to ensure that all areas are covered and that there is no duplication or double-counting.

## Briefing the Audit Teams

The audit teams are brought together and briefed. Several important questions are answered during this briefing: What business is the company in? What functions and tasks are performed by the employees in the buildings? Are there any unique features about the building layout or lighting systems? Are there any areas of the building that are exempt from the audit, such as a high-security R&D laboratory?

During the briefing, audit procedures are outlined, including individual duties and what teams will audit which areas, using the marked floor plans as guides. Boundary rules are reviewed and potential problems in getting reliable data are discussed. Any special conditions required by the client are reviewed, and a copy of these conditions is placed in each audit team's notebook. Special conditions usually address access and security, hours of availability, whom to call in case of trouble, and cleanup procedures.

The preprinted audit form is reviewed and as many copies as required are put in each team notebook, since one form is used for each space. Preprinted audit forms are shown in Chapter 10. Copies of the marked floor layouts are inserted in the front of the notebook.

Audit tools are checked and fresh batteries loaded in measuring instruments such as light meters and loggers. Personal safety equipment, such as hard hats, safety glasses, and earplugs (see Figure 4.1) are inspected, and company safety procedures are reviewed and a copy inserted in the front of the team notebook. Attention to safety issues is especially important for industrial surveys.

## Steps to a Successful On-Site Audit

When the audit team arrives at the survey site, a central location is selected as a control center. A laptop computer, loaded with a spreadsheet or an audit program such as LightPAD, is very useful for recording the data from each team while the teams are still on-site (see Chapter 6, "Commercial Software for Surveys and Analysis"). The preprinted audit forms are filled out (one for each space) as the audit teams move through the building. Keeping these forms in the same order as they are completed as the teams go through the building has proved valuable for filling in gaps in information discovered as the collected data is being entered into the computer. It is easier to revisit some areas while still on-site than it is to have to return to the site sometime later! Also, questions that arise during the audit can be directed to the client's representative and the equipment or condition can be viewed and discussed firsthand.

Walkie-talkies keep the audit teams in touch with the central location should the teams need assistance. Other tools that have proved valuable for recording information include small tape recorders, cameras, and video cameras. Because of security considerations, client companies need to be consulted on the use of cameras and video-cams in their facilities before the on-site visit. Allow several days for management approval for camera passes.

If building drawings are used, one member of the audit team carries the drawings and upon entering the space identifies if the space currently looks like the drawing. If it is different, space measurements are made using a measuring wheel or an ultrasonic ranging tool (see Figure 4.5) or by counting ceiling tiles. Laser ranging tools (see Figure 4.6) are available for larger areas and can be used for distances exceeding the maximum range of the ultrasonic units. For more information on these tools and how they are used to calculate space, refer to Chapter 4.

The team identifies the current tasks performed in the space they are evaluating and makes task illuminance measurements. At the same time, the team notes where task lighting is being used or where it might be needed. Each task (work plane) height is measured and recorded.

### Measuring Light Levels

A +/− 5% digital light meter (see Figures 4.9 and 4.10) is recommended to measure illuminance in footcandles (FC) or lux. A model with a hold feature that reads in 0.1 FC below 50 FC is preferred. The hold feature will allow readings to be made in situations where the digital display cannot be seen. Also, the additional precision of the autoranging model that reads in tenths of a FC below 50 FC is required when making reflectance measurements. Inexpensive taut-band analog meters (see Figure 4.9) are not recommended because they have low accuracy (as low as +/− 20%) and they do not read accurately after being bumped. Chapter 4 describes illuminance meters in detail.

Light levels are always measured at the task, which means that most measurements are made at a desk or bench top. Corridors are measured at floor level. Minimum and maximum readings are located and an average illuminance calculated. The techniques for measuring illuminance were described in detail in Chapter 4 (see also Chapter 8, Fact Sheet 13).

Daylight contributions are accounted for by first measuring with the electric lighting on and then measuring with the lighting off and then subtracting the two readings. This method eliminates any errors introduced by lighting systems' warming up to stable light output, which may occur if the lighting system is first turned off to measure the daylight contribution. If the daylight contribution is regarded by the auditor to be significant, this is noted for a separate investigation by the lighting professional as a potential daylighting control application.

### Collecting Lighting Data

Collecting lighting data requires assessing lighting quality, collecting fixture and lamp data, counting fixtures, determining the lighting load, and noting the condition of the fixtures and the spaces.

#### Lighting Quality

One way to improve the lighting in the spaces being audited is to improve the lighting quality. Lighting quality is assessed by learning to recognize poor lighting situations. Auditors record their subjective impressions, such as the appearance of light in the space. We have all been in spaces where it did not take a light meter to tell us that the space was either too dark or too bright. Proper illuminance, measured in footcandles or lux, is important for lighting quality and needs to be recorded.

To assess a glare situation, the "hat trick" and "salute" techniques are recommended. For the first technique, an auditor can wear a cap (with the bill turned to the front) and remove it while under the lighting system. If the cap bill is effective in shielding the glare, then when the cap is removed, the glare will be noticeable. To perform the second technique in lieu of the first, the auditor simply brings an extended hand to his or her (hatless) head, shielding the lighting with a "salute." Glare may be noticeable when the hand is taken away. If these techniques suggest glare, the visual comfort probability (VCP) can most likely be improved with fixture shielding.

In office desk or industrial bench lighting situations, being able to recognize veiling reflections is important. Veiling reflections are reflections of incident light superimposed on diffuse reflections that partially or totally obscure details by reducing task contrast. The effect is most severe when the light source is above and in front of the work surface. The best way to identify veiling reflections is with a test page such as the one shown in Figure 4.18.

Lamp flicker, produced by fluorescent and HID lamps operated on magnetic ballasts, can be eliminated or reduced to improve lighting quality and should be identified in the audit. A simple spinning top with special patterns can indicate flicker. A "flicker checker," available from Motorola Lighting (see Figure 4.8), may be used to identify magnetic ballasts and flicker.

### Describing Fixtures

Describing fixtures is best done by describing the nominal size in feet—although the word feet is usually omitted (1 × 4, 2 × 4, 2 × 2, etc.); the mounting (recessed, pendant, surface mount); the light control (lens or louver); whether it is static or air-handling (supply or return); and the number of lamps in each fixture. Pictures of fixture types, obtained from lighting handbooks or manufacturers' catalogs, can help the auditors identify various fixture types (Fact Sheet 6 in Chapter 8 outlines luminaire identification). Copies of these can be placed in the team notebooks for reference during the audit. Finding duplicate equipment (if available) in the building's electrical maintenance shop or storeroom is a way of identifying the fixture catalog number and is easier than using a ladder or lift to get to an installed lighting fixture.

### Counting Fixtures and Lamps (Bulbs)

Counting fixtures is not necessary when a building's drawings are kept up-to-date and show the current number of fixtures. The count can simply be obtained from the drawing. However, it is common to find that modifications have been made to the spaces but not to the drawings, so the auditors may have to count fixtures.

Counting fixtures is simple in regular ceiling layouts. Simply count the rows and columns and multiply. If an exact count is not required and spaces have a similar density of fixtures (measured as the area covered by each fixture, described in terms of square feet/fixture), a good estimate can be obtained by the following method. First, check a few rooms to estimate the average density, then apply this value to the spaces like those examined earlier to estimate the fixture count. The calculation is done by dividing the area of the space by the fixture density. For example, if each 2 × 4 fixture in an 8-foot-by-8-foot ceiling grid covers an area of 64 square feet, then in a space with an area of 700 square feet, the calculated estimate is 10 fixtures (700 divided by 64).

Some building lighting systems are arranged in bays. For this arrangement, count the number of fixtures in each bay and then count the number of bays and multiply. Watch for partial bays near the end of a run, usually near walls. Where different fixture types are found in the space, doing a separate count for each type is easiest (and more accurate), making a separate tour of the space for each fixture. A reset mechanical counter (see Figure 4.7) is useful when counting large quantities of fixtures. Remember to reset the counter when leaving each space.

Counting the number of lamps (bulbs) in each fixture is easy when the lamps can be seen. For those who wear eyeglasses, a trick makes it easy to see the number of fluorescent lamps if they are obscured by the lens of a direct fixture. Remove your glasses and invert them with the face of the eyeglass lens up. Look into the face of the eyeglass lens and tilt the glasses until the overhead fixture can be seen in the glasses. Distinct images of each bulb will be observed, making the bulbs easy to count. Commonly, the fixture will need to be opened to count the number of lamps. The quantity (or percentage) of burned-out lamps needs to be recorded on the audit form.

### Identifying Lamp (Bulb) Types

To identify lamp types in current use, look in nearby spaces or the electrical storeroom. In our experience, spare lamps are never far from where they are used. Interviewing the building staff people who replace lamps will help answer lamp questions. Fact Sheet 2 in Chapter 8 details lamp and bulb identification. Here are some special techniques to help identify lamps:

**Incandescent bulbs in downlights**: Use a 3 by 5 index card, punched with a small pinhole. The card will shield the direct glare from the bulb and allow a clear view of the identifying markings in the pinhole. Be aware that for bulbs with long accumulated burn hours, the identifying marks may have burned off and none can be seen!

**Lighted fluorescent tubes in opened fixtures**: Use photographers' gel, folded to provide the right amount of light transmission, and contrast to read the catalog markings at the end of the fluorescent tubes. It is common knowledge that in the first several fixtures examined, the tubes will be turned with their identifying end marking toward the fixture back and away from the viewer! Usually, several fixtures will need to be examined before one is found with the tubes turned so the identifying marks are readable.

**Lamps in high open bay fixtures**: These can sometimes be identified using binoculars. Turn off the lighting system before using binoculars so you are not blinded by the disability glare from the unshielded high-intensity discharge lamp. While the lamp's written identification cannot actually be read from a distance, even using binoculars, the lamp size and shape can be determined, as well as whether the bulb is clear or coated. This information, along with storeroom information, should be enough to identify the lamp type completely.

Sometimes the only way to identify lamps is to open selected fixtures. Do not overlook the fact that a mix of lamp types may be the prevailing condition in a facility. Note that there is a place provided on each audit form for the lamp (bulb) type.

### Determining Lighting Loads

Lighting loads in kW will be calculated in the analysis phase. To determine the lighting load, ballast data is needed. This information is obtained either by using the catalog number from the ballast label and looking up the ANSI test data in a ballast catalog or by using published data in books like the Green Lights *Lighting Upgrade Manual.*

The voltage of the lighting system can be verified from the ballast label. It is not a good idea to get the current (amps) information to calculate the watts per fixture by multiplying the volts times the amps. The power factor associated with magnetic ballasts can make enough difference between VA (volt-amps) and watts that it can introduce a calculation error.

Look in the electrical maintenance shop or storeroom for duplicate equipment. As mentioned previously, reading the ballast catalog number from the ballast label is easier to do on "downed" equipment than on installed lighting fixtures. Since the age of the ballast is usually stamped on the bottom of the ballast as a manufacturing date code, this information cannot be obtained without removing the ballast. If this information is needed to identify ballasts containing PCBs (polychlorinated biphenols) for disposal, other sources of this information may be easier to obtain. Dates of fixture purchase or maintenance records for ballast replacements, if available, can be used.

Even more likely than a mix of lamps is a mix of ballasts, and an estimate of the percentage of each type should be used to calculate the average watts per fixture. Lighting loads are calculated by multiplying the watts per fixture times the quantity of that fixture type. Many performance contractors now measure lighting systems with power meters, taking measurements at each lighting distribution panel. The main advantage of this method is it is the most accurate way of determining lighting loads; also, it eliminates the problems of estimating the mix of lamps and ballast types and the use of ANSI test values. The disadvantage of this method in comparison to using ballast data is that it takes more time and uses more expensive measuring equipment.

### Determining the Condition of Fixtures and Spaces

The condition of fixtures and spaces is dependent upon the state of maintenance in a facility. A determination of the state of maintenance is valuable information because in many cases, lack of maintenance can reduce the potential of a lighting upgrade (see Chapter 8, Fact Sheet 14). It is easy to discover a facility's level of attention to maintenance. Unlighted fixtures, old lamps (evidenced by dark ends or low light output), dirty or rusty fixtures with broken lenses or dirty louvers, and dirty ceilings or walls are all evidence of maintenance inactivity. Dirt tracks leading to air return fixtures suggest that HVAC filters are not being replaced at regular intervals, evidence of a lack of attention to maintenance in general. A checklist of maintenance conditions for the fixtures and the wall, ceiling, and floor surfaces is found in an area of the preprinted audit form to make it easy for the auditors to simply describe these conditions.

## Collecting Other Data

Other data that need to be collected include lighting system voltages, switching methods and zoning, and hours of use.

### Lighting System Voltage

Measuring the applied line voltages can help identify lighting system problems due to abnormally high or low voltages. A handy tool for auditors to carry is a plug-in expanded-scale voltmeter like that shown in Figure 4.12. Plugging this meter into outlets does not measure the lighting system voltage directly, but the meter can be used as a screening device to indicate when measuring the lighting system voltage directly at lighting panels may be appropriate. If the auditor finds that the outlet voltage is out of tolerance (as shown outside the green band on the meter), then the lighting system voltages should be measured directly at the lighting distribution panels.

### Switching Methods and Zoning

Switching methods and zoning information is recorded for use later when redesigning the controls or applying occupancy sensors. A checklist of controls that includes breaker, wall switch, pushbutton/contactor controls, and timer choices are included on the preprinted audit form. Other switching arrangements, such as in-board/out-board switching of three- or four-lamp fluorescent fixtures, are noted in the written comments. Zoning is usually described by the number of fixtures on a control.

### Hours-of-Use

The actual amount of time lighting systems are in use in a building is the most difficult information to obtain. Building personnel, when interviewed, may explain that the lights are always turned off at night, but a midnight drive by the facility will often show evidence to the contrary. The best way to obtain hours-of-use information is to monitor the systems with a logger. Many loggers are available today, and they vary widely in their sophistication and price. Chapter 4 describes lighting loggers and details of their operation.

The lighting of unoccupied spaces wastes energy. The occupancy of spaces can be monitored to identify this waste by means of several techniques that vary in both cost and accuracy. There are special loggers that have an integral occupancy sensor or have inputs from a portable sensor to evaluate the potential savings from occupancy sensors. The least expensive yet most accurate method is observation during the on-site audit. When the audit is performed during normal occupancy it is common to observe conference rooms with no meetings or small offices with no one in them.

After the hours-of-use information is obtained, it is a simple matter to calculate the lighting energy in kilowatt-hours (kWh) by multiplying the load of each system, in kW, by the hours of use for that system.

### Conducting Interviews

Interviewing key facilities personnel will help bridge gaps in the collected data, especially when lamp (bulb) data is in doubt. Facility engineers and electricians are the best source of this information. Resist the temptation to show off your lighting knowledge and instead concentrate on asking questions in a simple, direct, nonthreatening way.

### Auditing Exterior Lighting Systems

Parking lot lights, facade floodlighting, building-mounted fixtures, and walkway lighting are included in the exterior lighting systems audit. This part of the audit begins at dusk so the auditor can observe if fixtures are controlled by photocells. Each fixture type is counted, any high-pressure sodium lamp cycling is noted, and dark areas are described.

### Auditing the Auditors

Spot checks are made during the on-site audit by the audit team leader to ensure that all areas are covered, that there is no double-counting, and that reliable identification and counts are being obtained. If any data is in doubt, the area in doubt is audited again, usually by another team. During break times, the auditors meet at the control central to review progress, ask questions, and discuss problem areas.

### Completing the Audit

When all areas have been covered, team notebooks are reviewed for completeness, and tools and safety equipment are inventoried and packed. A thorough cleanup of the area used as a control center is important to maintain good client relations.

## Summary

On-site data collection is the heart of the survey or audit, because without reliable data, the other steps in the audit process will be flawed. Auditors who take the time and effort to collect reliable and relevant data will avoid costly mistakes. The goal of an on-site audit is to obtain solid, accurate information in a short time. The time spent collecting the right information in sufficient quantity (and no more) is the key to an effective audit.

Some building information needs to be collected and reviewed before the auditing team goes to the building site. The correct electric rate schedule and electric bills for the previous 12 months are needed, as well as building electrical drawings to allow for a preview of the building lighting systems.

An organized plan, prepared prior to the on-site visit, can reduce on-site time. The plan should outline each auditor's task and be tailored to the age, size, and type of building being surveyed and to the type of lighting systems in the building. On-site audits must be done one step at a time, in an appropriate order. Audit teams should be briefed and their questions answered before the on-site survey or audit begins. The practical tips and techniques discussed in the chapter will help the auditor collect data effectively. Interviewing key facilities personnel can help bridge gaps in the collected data, especially when lamp (bulb) data is in doubt facility engineers and electricians are the best source for this information.

# Chapter 6
# Commercial Software for Surveys and Analysis

There are many benefits that can be derived from using commercial software for surveys and analysis. The data is collected in a systematic way, providing a faster, more accurate, more complete and understandable survey. Using software can reduce the total audit process time by reducing the time spent in transferring or repeating data from the data collection phase to the analysis phase. Programs provide a database for management reports and facility models can be generated by creating a library of audits categorized according to building type. Information is readily available for follow-up visits and evaluations. Representative utility cost and energy conservation measures (ECM) analyses can easily be performed and cataloged for various building types. In addition, using software programs eliminates the potential inaccuracy and tedious hours of manually performing the calculations required for analyzing various retrofit options.

The disadvantages of using computer programs include the oftentimes considerable data entry. Also, laptops can be cumbersome to use in the field, and errors in input can result in erroneous results (this is known as GIGO: "garbage in, garbage out"). The smaller size and weight of PDAs (personal digital assistants) and palm tops will be appreciated by auditors everywhere. The computer programs now available are easy to use and will usually prove valuable to the auditor. The most useful audit and analysis software is described in the remainder of this chapter.

## Computer Spreadsheets

Programs such as Lotus 1-2-3 or EXCEL can be used to set up spreadsheets for either data collection or analysis functions. A simple spreadsheet for data collection is shown in Figure 6.1. It is an actual spreadsheet used for a large nursing home in southern California. This spreadsheet was created from a walk-thru survey and shows existing fixture types, the lamps used in each fixture type, and the estimated watts per fixture. The spreadsheet is also used for the analysis of each retrofit recommended, with data entered for the new wattage, the calculated power reduction for each fixture type, the estimated hours-of-use data, a calculation of annual kWh savings, the estimated annual savings per fixture, the estimated material cost entries, and a calculation of simple payback. Note that there is no entry for fixture quantities. Fixture quantities would usually be estimated for a walk-thru survey and each fixture would be counted in an audit. In this case, the client desired to know, on a fixture basis, the retrofit potential and where to start. Only material costs were used since the original idea was to use in-house labor for the installation. A survey spreadsheet is easy to set up for those familiar with computer spreadsheets and can be tailored for each client's needs.

Figure 6.2 shows a spreadsheet used for analysis. This spreadsheet is one sheet of a large EPA Green Lights audit performed for a large Midwest manufacturing company. This spreadsheet was used to compile the locations for all fixtures of a given type: 4-lamp, 2 × 4 recessed troffers. Each row provides information on a particular room or small area. In the columns are recorded the area of each space, the fixture quantity, the fixture type, the fixture size, the lamp and ballast types (using the Green Lights codes), the lamp watts, the lamps per fixture, and the fixture watts. The total number of watts is calculated by inserting a formula that multiplies the fixture quantity times the fixture watts. Annual hours of use for each space is entered, and the annual kWh is calculated by inserting a formula that multiplies the total watts times the annual hours of use and dividing by 1000 (to convert watts to kW). In the last row, totals are calculated for all the rows, using the @sum function for the total area, the total fixture quantity, the total watts and kW, and the total annual kWh.

A spreadsheet template is a prepared form with data blanks and formulas. The user can start entering data as soon as the template is loaded on a PC. The idEA$ program template, named after the Alabama state energy service of the same name, can be used to compare up to three lighting upgrade alternatives to an existing system. The idEA$ template was developed at the Johnson Research Center at the University of Alabama in Huntsville by T. Kenneth Spain, P.E., CEM, CLEP, to support walk-thru audits. This spreadsheet provides the essential comparative data for evaluating the cost-effectiveness of commercial lighting retrofits. The impact of the alternatives on environmental emissions is also calculated to help communicate external costs and sell the project. Avoided environmental emissions are

Sunny Dale Lutheran Home
22445 Cupateana Rd.
Cupateana, CA
Lighting Survey
Performed April 19, 1995 & 1/23/96
by Effective Lighting Solutions, Inc.
Columbus, Ohio 43205
Survey conducted by John Fetters
Certified Lighting Efficiency Professional

| Line | Priority | Fix Type | Existing Light Fixture Description | Est Watts | Recommended Change | Watts | Power Reduction | Est hrs per year | Annual kWh Savings | Estimated Annual Savings/Fix | Estimated Material Cost | Simple Payback (yrs) | Comments |
|---|---|---|---|---|---|---|---|---|---|---|---|---|---|
| 1 | 1 | 5 | Round, surface mount, w/FC12T9 circleline lp | 40 | ALTA SMA-38 w/2-D lamp & ELB | 38 | 2 | 8,760 | 18 | $1.58 | $75.00 | 47.6 | Area too dark now; est 50% more light - DO for Safety! |
| 2 | 2 | 1 | Round, surface mount, w/32/22 circleline lps | 70 | ALTA SMA-38 w/2-D lamp & ELB | 38 | 32 | 8,760 | 280 | $25.23 | $75.00 | 3.0 | Equivalent light |
| 3 | 1 | 8 | Various fixtures, w/75R30 lamps | 75 | 50w T-H PAR30 flood (50PAR30LN/CAP/WFL) | 50 | 25 | 8,760 | 219 | $19.71 | $7.00 | 0.4 | Low cost, short payback, high priority (EPAct) |
| 4 | 1 | 3 | 4', 4-lamp, w/F40CW lamps | 190 | 4-T-8 lamps + electronic ballast (ELB) | 106 | 84 | 8,760 | 736 | $66.23 | $30.00 | 0.5 | Long burn hours = short payback = high priority (EPAct) |
| 5 | 1 | 14 | Cove lights, 2-lamp, w/F40CW lamps | 95 | 2-T-8 lamps + electronic ballast (ELB) | 60 | 35 | 8,760 | 307 | $27.59 | $23.00 | 0.8 | Long burn hours = short payback = high priority (EPAct) |
| 6 | 1 | 18 | 4-chandeliers w/4-G lamps (Ruud Comm Ctr) | 1,600 | 15 watt Compact fluorescent G40 lamp | 240 | 1,360 | 3,000 | 4,080 | $367.20 | $320.00 | 0.9 | Medium cost, less than 1 yr payback |
| 7 | 1 | 9 | Various fixtures w/75 w, incand A lamp | 75 | SLS-20 integral CFL | 20 | 55 | 4,000 | 220 | $19.80 | $20.00 | 1.0 | Low cost, short payback, high priority |
| 8 | 1 | 7 | Surface mtd socket, w/G40 incand lamp | 100 | 15 watt Compact fluorescent G40 lamp | 15 | 85 | 2,200 | 187 | $16.83 | $20.00 | 1.2 | Low cost, short payback, high priority |
| 9 | 1 | 10 | Various fixtures w/60 w, incand A lamp | 60 | 42 w capsulite, tungsten-halogen lamp | 20 | 40 | 4,000 | 160 | $14.40 | $20.00 | 1.4 | Low cost, short payback, high priority |
| 10 | 1 | 16 | 4x4, 6-lamp, w/F40CW lamps | 285 | 6-T-8 retrofit lamps + 1-2 lp & 1-4 lp ELB | 180 | 105 | 4,000 | 420 | $37.80 | $55.00 | 1.5 | Medium cost, less than 2 yr payback (EPAct) |
| 11 | 1 | 12 | Beauty shop w/21-G lamps beside mirrors | 2,100 | 15 watt Compact fluorescent G40 lamp | 315 | 1,785 | 1,560 | 2,785 | $250.61 | $420.00 | 1.7 | Complaint of "too hot" from incand G-lamps |
| 12 | 2 | 15 | 2x2, 2-lamp, w/FB40CW U lamps | 95 | 2-T-8 retrofit lamps + ELB | 60 | 35 | 4,000 | 140 | $12.60 | $29.00 | 2.3 | EPAct affected lamp - October 31,95 deadline |
| 13 | 2 | 19 | Exit light w/2-20T6-1/2 incand lamps | 40 | New Conservation Alliance LED exit lights | 2 | 38 | 8,760 | 333 | $29.96 | $68.00 | 2.3 | Energy savings and maintenance savings |
| 14 | 2 | 2 | 4', 2-lamp, w/F40CW lamps - std ballast | 95 | 2-T-8 lamps + electronic ballast (ELB) | 60 | 35 | 3,000 | 105 | $9.45 | $23.00 | 2.4 | EPAct affected lamp - October 31,95 deadline |
| 15 | 2 | 13 | 2x4, 2-lamp, w/F40CW lamps | 95 | 2-T-8 lamps + electronic ballast (ELB) | 60 | 35 | 3,000 | 105 | $9.45 | $23.00 | 2.4 | EPAct affected lamp - October 31,95 deadline |
| 16 | 3 | 4 | 8', 2-lamp, w/F96WM lamps | 140 | 4-T-8 lamps + ELB + Harris conversion kit | 106 | 34 | 8,760 | 298 | $26.81 | $90.00 | 3.4 | High cost, small savings = low priority |
| 17 | 3 | 11 | Residence baths w/3-G lamps over sinks | 180 | 11 watt Compact fluorescent G30 lamp | 33 | 147 | 550 | 81 | $7.28 | $60.00 | 8.2 | Short burn hours = long payback = low priority |
| 18 | 3 | 17 | Surface mtd, w/FC6T9 circleline fluorescent | 28 | New fixture w/13 w TT CFL lp (damp rated) | 17 | 11 | 4,380 | 48 | $4.34 | $35.00 | 8.1 | High cost, long payback, do after others done |
| 19 | 3 | 20 | Exit light w/2-F6T5 fluor lamps | 20 | New LED exit lights | 3 | 17 | 8,760 | 149 | $13.40 | $70.00 | 5.2 | Phase in as units are replaced to lower maint costs |
| 20 | 3 | 21 | Exit light w/AC/DC lamps | 36 | New LED exit lights | 3 | 33 | 8,760 | 289 | $26.02 | $130.00 | 5.0 | Phase in as units are replaced to lower maint costs |
| | | 6 | Recessed, square, w/FC8T9 circleline lamp | 30 | None at this time | 30 | 0 | | 0 | $0.00 | $0.00 | | |

**FIGURE 6.1** Simple spreadsheet used for data collection.

Refrigrus Co.
Novaksville, IN
1994 LIGHTING SURVEY
4-Lp 2x4 Recessed Troffers

| LOCATION | SPACE | NET SF | QTY Fix | FIX TYPE | FIX SIZE | LAMP TYPE | LAMP DESCRIP | LAMP WATTS | LPS/ FIX | BALLAST TYPE | LPS/ BAL | FIXTURE WATTS | TOTAL WATTS | ANNUAL HRS USE | ANNUAL KWH | WATTS/ SF |
|---|---|---|---|---|---|---|---|---|---|---|---|---|---|---|---|---|
| Building 1 | stairwell | 200.0 | 5 | 15 | 50 | 56 | RW | 34 | 4 | 80 | 2 | 164 | 820 | 4,200 | 3,444 | 4.1 < |
| | Toth | 400.0 | 6 | 15 | 50 | 56 | RW | 34 | 4 | 80 | 2 | 164 | 984 | 4,200 | 4,133 | 2.5 |
| | Blakey | 288.0 | 6 | 15 | 50 | 56 | RW | 34 | 4 | 80 | 2 | 164 | 984 | 4,200 | 4,133 | 3.4 |
| | Sec Off-Lynch | 138.0 | 2 | 15 | 50 | 59 | STD | 40 | 4 | 80 | 2 | 192 | 384 | 4,200 | 1,613 | 2.8 |
| | Exec Conf Rm | 528.0 | 6 | 15 | 50 | 59 | STD | 40 | 4 | 80 | 2 | 192 | 1,152 | 4,200 | 4,838 | 2.2 |
| | Offutt | 400.0 | 8 | 15 | 50 | 59 | STD | 40 | 4 | 80 | 2 | 192 | 1,536 | 4,200 | 6,451 | 3.8 |
| | Boot | 224.0 | 4 | 15 | 49 | 59 | STD | 20 | 4 | 80 | 2 | 112 | 448 | 4,200 | 1,882 | 2.0 |
| | | | 2 | 15 | 50 | 59 | STD | 40 | 4 | 80 | 2 | 192 | 384 | 4,200 | 1,613 | |
| | Secretary | 310.0 | 6 | 15 | 50 | 59 | STD | 40 | 4 | 80 | 2 | 192 | 1,152 | 4,200 | 4,838 | 3.7 |
| | corridor | 192.0 | 2 | 15 | 50 | 59 | STD | 40 | 4 | 80 | 2 | 192 | 384 | 4,200 | 1,613 | 2.0 |
| | | | 1 | 15 | 50 | 56 | RW | 34 | 4 | 80 | 2 | 164 | 164 | 8,760 | 1,437 | |
| | Miller | 140.0 | 5 | 15 | 50 | 56 | RW | 34 | 4 | 80 | 2 | 164 | 820 | 4,200 | 3,444 | 5.9 < |
| | hall to personnel | 160.0 | 4 | 15 | 50 | 56 | RW | 34 | 4 | 80 | 2 | 164 | 656 | 4,200 | 2,755 | 4.1 < |
| | personnel | 1,736.0 | 46 | 15 | 50 | 56 | RW | 34 | 4 | 80 | 2 | 164 | 7,544 | 4,200 | 31,685 | 4.3 < |
| | alcove to pers | 185.0 | 4 | 15 | 50 | 56 | RW | 34 | 4 | 80 | 2 | 164 | 656 | 4,200 | 2,755 | 3.5 |
| | storage-personn | 80.0 | 3 | 15 | 50 | 56 | RW | 34 | 4 | 80 | 2 | 164 | 492 | 4,200 | 2,066 | 6.2 < |
| | Jillson | 144.0 | 4 | 15 | 50 | 56 | RW | 34 | 4 | 80 | 2 | 164 | 656 | 4,200 | 2,755 | 4.6 < |
| | supply rm | 96.0 | 3 | 15 | 50 | 56 | RW | 34 | 4 | 80 | 2 | 164 | 492 | 4,200 | 2,066 | 5.1 < |
| | Weldon | 120.0 | 4 | 15 | 50 | 56 | RW | 34 | 4 | 80 | 2 | 164 | 656 | 4,200 | 2,755 | 5.5 < |
| | C. Smith | 120.0 | 4 | 15 | 50 | 56 | RW | 34 | 4 | 80 | 2 | 164 | 656 | 4,200 | 2,755 | 5.5 < |
| | Spencer | 80.0 | 2 | 15 | 50 | 56 | RW | 34 | 4 | 80 | 2 | 164 | 328 | 4,200 | 1,378 | 4.1 < |
| | personn conf rm | 168.0 | 6 | 15 | 50 | 56 | RW | 34 | 4 | 80 | 2 | 164 | 984 | 4,200 | 4,133 | 5.9 < |
| | Springer | 168.0 | 6 | 15 | 50 | 56 | RW | 34 | 4 | 80 | 2 | 164 | 984 | 4,200 | 4,133 | 5.9 < |
| | next to Springer | 168.0 | 6 | 15 | 50 | 56 | RW | 34 | 4 | 80 | 2 | 164 | 984 | 4,200 | 4,133 | 5.9 < |
| | Castrali | 168.0 | 6 | 15 | 50 | 56 | RW | 34 | 4 | 80 | 2 | 164 | 984 | 4,200 | 4,133 | 5.9 < |
| | Schmitt | 168.0 | 6 | 15 | 50 | 56 | RW | 34 | 4 | 80 | 2 | 164 | 984 | 4,200 | 4,133 | 5.9 < |
| | physical assess | 90.0 | 1 | 15 | 50 | 56 | RW | 34 | 4 | 80 | 2 | 164 | 164 | 4,200 | 689 | 1.8 |
| | first aid | 80.0 | 2 | 15 | 50 | 56 | RW | 34 | 4 | 80 | 2 | 164 | 328 | 4,200 | 1,378 | 4.1 < |
| | x-ray | 80.0 | 2 | 15 | 50 | 56 | RW | 34 | 4 | 80 | 2 | 164 | 328 | 4,200 | 1,378 | 4.1 < |
| | drug screening | 75.0 | 2 | 15 | 50 | 56 | RW | 34 | 4 | 80 | 2 | 164 | 328 | 4,200 | 1,378 | 4.4 < |
| | first-aid hall | 200.0 | 5 | 15 | 50 | 56 | RW | 34 | 4 | 80 | 2 | 164 | 820 | 4,200 | 3,444 | 4.1 < |
| | first-aid lobby | 112.0 | 3 | 15 | 50 | 56 | RW | 34 | 4 | 80 | 2 | 164 | 492 | 4,200 | 2,066 | 4.4 < |
| | first-aid office | 90.0 | 2 | 15 | 50 | 56 | RW | 34 | 4 | 80 | 2 | 164 | 328 | 4,200 | 1,378 | 3.6 |
| | nurse's office | 104.0 | 2 | 15 | 50 | 56 | RW | 34 | 4 | 80 | 2 | 164 | 328 | 4,200 | 1,378 | 3.2 |
| | ER | 192.0 | 4 | 15 | 50 | 56 | RW | 34 | 4 | 80 | 2 | 164 | 656 | 4,200 | 2,755 | 3.4 |
| | nurse's station | 104.0 | 2 | 15 | 50 | 56 | RW | 34 | 4 | 80 | 2 | 164 | 328 | 4,200 | 1,378 | 3.2 |
| | file rm | 144.0 | 4 | 15 | 50 | 56 | RW | 34 | 4 | 80 | 2 | 164 | 656 | 4,200 | 2,755 | 4.6 < |
| | office & alcove | 128.0 | 5 | 15 | 50 | 56 | RW | 34 | 4 | 80 | 2 | 164 | 820 | 4,200 | 3,444 | 6.4 < |
| | indust rel's | 132.0 | 4 | 15 | 50 | 56 | RW | 34 | 4 | 80 | 2 | 164 | 656 | 4,200 | 2,755 | 5.0 < |
| | off adj to 1st aid | 121.0 | 3 | 15 | 50 | 56 | RW | 34 | 4 | 80 | 2 | 164 | 492 | 4,200 | 2,066 | 4.1 < |
| | corridor | 192.0 | 6 | 15 | 50 | 56 | RW | 34 | 4 | 80 | 2 | 164 | 984 | 4,200 | 4,133 | 5.1 < |
| | lobby | | 9 | 15 | 50 | 56 | RW | 34 | 4 | 80 | 2 | 164 | 1,476 | 4,200 | 6,199 | |
| | hall | 60.0 | 1 | 15 | 50 | 56 | RW | 34 | 4 | 80 | 2 | 164 | 164 | 4,200 | 689 | 2.7 |
| | meeting rm | 135.0 | 2 | 15 | 50 | 56 | RW | 34 | 4 | 80 | 2 | 164 | 328 | 4,200 | 1,378 | 2.4 |
| | receptionist | 256.0 | 4 | 15 | 50 | 56 | RW | 34 | 4 | 80 | 2 | 164 | 656 | 4,200 | 2,755 | 2.6 |
| TOTALS | S = | 8,676 | 220 | | | | | | | | | Watts = | 36,600 | kWh= | 154,468 | 4.2 |
| | | | | | | | | | | | | kW= | 36.6 | | | Avg^ |

**FIGURE 6.2** Spreadsheet used for analysis as part of a large EPA Green Lights audit.

calculated for carbon dioxide, sulfur dioxide, and nitrous oxides. HVAC impacts and a choice of average costs or demand-incremental electrical rates add sophistication to the template. In its latest version, the spreadsheet also provides a simplified life-cycle costing of each lighting alternative. Both total life-cycle cost (LCC) and normalized LCC are provided. Simple payback, simple investor's rate of return, discounted payback, the present value of the net savings, the savings-to-investment ratio (SIR), and the adjusted internal rate of return are calculated and displayed in the "Comparative Statistics" section at the end of the spreadsheet. Table 6.1 shows a printout of the input data file, filled in for a classroom lighting application. Table 6.2 shows the results of the calculations performed by the spreadsheet program. An important statistic for determining retrofit feasibility is given in the power density row.

## IESNA Lighting Handbook Software

Another spreadsheet template, Lighting Handbook Software, is available from the Illuminating Engineering Society of North America (IESNA). Two formats are provided on the 3.5-inch disk, one for

**TABLE 6.1**

Filename: LIGHT3.WK1
Project: Classroom
Description: 24' × 24' × 10' classroom with nine 2' × 4' 4-tube troffers
Continue spot replacement of lamps

| Input Data | Existing | Proposed 1 | Proposed 2 | Proposed 3 |
|---|---|---|---|---|
| **Fixture Characteristics** | | | | |
| Fixture Type | 2 × 4-4L | 2 × 4-4L | 2 × 4-2L-SR | 2 × 4-2L-SR |
| Number of Fixtures | 9 | 9 | 9 | 9 |
| Fixture Power, W/fixture | 190 | 105 | 95 | 62 |
| Coefficient of Utilization (CU) | 0.50 | 0.50 | 0.75 | 0.75 |
| Estimated Useful Life, yrs | 15 | 15 | 15 | 15 |
| **Lamp Characteristics** | | | | |
| Lamp Type | F40T12 | F32T8 | F40T12 | F32T8 |
| Lamps per Fixture | 4 | 4 | 2 | 2 |
| Average Rated Lamp Life, hrs | 20,000 | 15,000 | 20,000 | 20,000 |
| Lamp Output, lumens (initial) | 3200 | 3050 | 3200 | 2900 |
| **Light Loss Factors** | | | | |
| Ballast Factor | 0.94 | 0.95 | 0.94 | 0.88 |
| Other Nonrecoverable | 1.00 | 1.00 | 1.00 | 1.00 |
| Lamp Burnout Factor | 1.00 | 1.00 | 1.00 | 1.00 |
| Lamp Lumen Depreciation | 0.75 | 0.90 | 0.90 | 0.90 |
| Luminaire Dirt Depreciation | 0.90 | 0.90 | 0.90 | 0.90 |
| Room Surface Dirt Depreciation | 1.00 | 1.00 | 1.00 | 1.00 |
| **Situation** | | | | |
| Daily Operation, hrs/day | 12 | 12 | 12 | 12 |
| Weekly Operation, days/wk | 5 | 5 | 5 | 5 |
| Annual Operation, wks/yr | 50 | 50 | 50 | 50 |
| Area Served, square feet | 576 | 576 | 576 | 576 |
| Measured Light Level, FC | 92 | 97 | 78 | 64 |
| **HVAC Impact** | | | | |
| Fraction of Annual Lighting Heat to Cooling | 0.53 | 0.53 | 0.53 | 0.53 |
| Cooling Seasonal Efficiency, MCOP | 3 | 3 | 3 | 3 |
| Fraction of Annual Lighting Heat to Heating | 0.11 | 0.11 | 0.11 | 0.11 |
| Heating Seasonal Efficiency, % | 75 | 75 | 75 | 75 |
| **Installation and Replacement** | | | | |
| Fixed Installation Cost, $ | 0 | 0 | 0 | 0 |
| Variable Installation Cost, $/fixture | 0 | 57 | 62 | 101 |
| Replacement Lamp Cost, $/lamp | 0.0302 | 0.0302 | 0.0302 | 0.0302 |
| Replacement Labor, hrs/lamp | 0.25 | 0.25 | 0.25 | 0.25 |
| Labor Cpst. $/hr | 10 | 10 | 10 | 10 |
| **Energy** | | | | |
| Energy Cost Method (0 = avg, 1 = incr) | 1 | 1 | 1 | 1 |
| Average Energy Cost, $/kWh | $0.0700 | $0.700 | $0.700 | $0.700 |
| Incremental Energy Cost, $/kWh | $0.0302 | $0.0302 | $0.0302 | $0.0302 |
| Incremental Demand Cost, $/kW/month | $8.10 | $8.10 | $8.10 | $8.10 |
| Heating Energy Cost, $/MBTU | $5 | $5 | $5 | $5 |
| **Financial** | | | | |
| Incremental Income Tax Rate, % | 0 | 0 | 0 | 0 |
| Depreciation Rate, %/yr | 0 | 0 | 0 | 0 |
| Discount Rate, % | 4 | 4 | 4 | 4 |
| Uniform Present Value Factor (UPV or UPV*) | 11.12 | 11.12 | 11.012 | 11.12 |

Lotus 1-2-3 and the other for EXCEL. Up to 10 lighting systems can be compared in the Life-Cycle Cost-Benefit Analysis Spreadsheet for Lighting Systems. Figure 6.3 shows an example of the spreadsheet that compares fluorescent and HID industrial systems.

## SYLVIA for PC Windows

The latest version of SYLVIA PC, from OSRAM-SYLVANIA, is now a Windows-based software offering that consists of five analysis models, that use data obtained from a lighting survey. These five models allow the user to quickly and accurately select and maintain the most efficient, effective lamps and lighting systems available today. This version of SYLVIA PC is capable of calculating rebates and avoided environmental emissions to provide a complete description of costs, advantages, and return on investment.

The Relight Energy Analysis (RENA) model computes savings in energy costs associated with

**TABLE 6.2**

Filename: LIGHT3.WK1
Project: Classroom
Description: 24' × 24' × 10' Classroom with nine 2' × 4' 4-tube troffers
Continue spot replacement of lamps

| Calculated Data | | Existing | Proposed 1 | Proposed 2 | Proposed 3 |
|---|---|---|---|---|---|
| **Light Quantity** | | | | | |
| Total Initial Light Output, lumens | | 115200 | 109800 | 57600 | 52200 |
| Total Nonrecoverable Light Loss Factor | | 0.94 | 0.95 | 0.94 | 0.88 |
| Modified Initial Light Output, lumens | | 108288 | 104310 | 54144 | 45936 |
| Estimated Initial Illuminance, FC | | 94 | 91 | 71 | 60 |
| Total Recoverable Light Loss Factor | | 0.68 | 0.81 | 0.68 | 0.81 |
| Total Maintained Light Output, lumens | | 73094 | 84491 | 36547 | 37208 |
| Estimated Maintained Illuminance, FC | | 63 | 73 | 48 | 48 |
| **Situation** | | | | | |
| Total Connected Load, kW | | 1.710 | 0.0945 | 0.855 | 0.558 |
| Annual Operation, hrs/yr | | 3000 | 3000 | 3000 | 3000 |
| Power Density, W/sf | | 2.97 | 1.64 | 1.48 | 0.97 |
| System Efficacy, maintained lumens/W | | 42.75 | 89.41 | 42.75 | 66.68 |
| **Installation and Replacement** | | | | | |
| Total Installed Cost, $ | | 0 | 513 | 558 | 909 |
| Average Lamp Replacements, lamps/yr | | 5.4 | 7.2 | 2.7 | 2.7 |
| Replacement Costs, $/lamp | | 3.50 | 5.50 | 3.50 | 5.50 |
| Total Annual Replacement Cost, $ | | 18.90 | 39.60 | 9.45 | 14.85 |
| **Energy** | | | | | |
| Lighting Energy Consumption, kWh/yr | | 5130 | 2835 | 2565 | 1674 |
| Lighting-Related Cooling Energy, kWh/yr | | 906 | 501 | 453 | 296 |
| Total Electric Energy Consumption, kWh/yr | | 6036 | 3336 | 3018 | 1970 |
| Annual Electric Energy Cost (incr), $ | | 348.63 | 192.66 | 174.31 | 113.76 |
| Lighting-Displaced Heating Energy, MBTU/yr | | 2.57 | 1.42 | 1.28 | 0.84 |
| Lighting-Displaced Heating Savings, $/yr | | 12.84 | 7.1 | 6.42 | 4.19 |
| Net Annual Energy Cost, $/yr | | 335.79 | 185.57 | 167.89 | 109.57 |
| **Financial** | | | | | |
| Before-Tax Operating Cost, $/yr | | $354.69 | $225.17 | $177.34 | $124.42 |
| Depreciation, $/yr | | $0 | $0 | $0 | $0 |
| Net Influence on Taxable Income, $/yr | | $354.69 | $225.17 | $177.34 | $124.42 |
| After-Tax Cash Flow, $/yr | | $354.69 | $225.17 | $177.34 | $124.42 |
| After-Tax Cash Savings, $/yr | | $0 | $129.52 | $177.34 | $230.27 |
| **Environmental Emissions** | **lbs/kWh** | | | | |
| Carbon Dioxide, lbs/yr | 1.48 | 8933.7 | 4937.1 | 4466.9 | 2915.2 |
| Sulfur Dioxide, lbs/yr | 0.016 | 96.6 | 53.4 | 48.3 | 31.5 |
| Nitrous Oxides, lbs/yr | 0.007 | 42.3 | 23.4 | 21.1 | 13.8 |
| **Life-Cycle Costs** | | | | | |
| Total Life-Cycle Cost (LCC), PV$ | | $3944 | $3017 | $2530 | $2292 |
| Normalized LCC, PV$/maintained FC | | $62.15 | $41.13 | $53.16 | $47.32 |
| **Comparative Statistics (vs. Existing)** | | | | | |
| Maintained Illuminance Change, % | | | 15.6 | -25.0 | -23.6 |
| Simple Payback, yrs | | | 4.0 | 3.1 | 3.9 |
| Simple (Investor's) Rate of Return, % | | | 25.2 | 31.8 | 25.3 |
| Discounted Payback, yrs | | | 4.4 | 3.4 | 4.4 |
| Net Savings (NS), PV$ | | | $927 | $1414 | $1651 |
| Savings-to-Investment Ratio (SIR) | | | 2.81 | 3.53 | 2.82 |
| Adjusted Internal Rate of Return (AIRR), % | | | 11.4 | 13.1 | 11.4 |

| PRELIMINARIES | | |
|---|---|---|
| Area of installation (sq. ft.): | 11000 | |
| Owner's cost of capital (decimal fraction): | 8% | |
| Owner's income tax rate (decimal fraction): | 25% | |
| Years in planning horizon or economic life: | 10 | |
| | | |
| A. INITIAL COSTS | System 1 | System 2 |
| 1. Lighting system installed cost: | $0 | $18,000 |
| Number of lamps in system: | 260 | 36 |
| Lamp wattage: | 60 | 400 |
| Ballast wattage per lamp: | 10 | 28 |
| 2. Total kW lighting: | 18.20 | 15.41 |
| 3. Tons of air conditioning required to dissipate heat of lighting: | 5.18 | 4.38 |
| Cost per ton of air conditioning equipment: | | $0 |
| 4. First cost of air conditioning tons in line A.3: | $0 | $0 |
| 5. Reduction in first cost of heating equipment (enter as negative): | | |
| 6. Utility rebates (enter as negative): | $0 | $0 |
| 7. Other first costs engendered by the presence of lighting system: | | |
| 8. Subtotal mechanical and electrical installed cost: | $0 | $18,000 |
| 9. Initial taxes | | |
| 10. Total costs | $0 | $18,000 |
| 11. Installed cost per square foot (memo): | $0.00 | $1.64 |
| 12. Watts of lighting per square foot (memo): | 1.65 | 1.40 |
| 13. Residual value at end of economic life: | | |
| | | |
| B. ANNUAL POWER AND MAINTENANCE COSTS | | |
| Operating hours | 8760 | 8760 |
| Energy cost per kWh (in dollars): | $0.05 | $0.05 |
| 1. Lighting energy cost: | $7,972 | $6,749 |
| Air conditioning operating hours: | 5600 | 5600 |
| Air conditioning efficiency (kW/ton): | 1 | 1 |
| 2. Air conditioning energy cost: | $1,449 | $1,227 |
| Annual maintenance cost per ton of installed air conditioning: | $0 | $0 |
| 3. Air conditioning maintenance: | $0 | $0 |
| Cost of heating fuel in dollars per MBtu | $5.50 | $5.50 |
| 4. Reduction in heating cost: | ($631) | ($534) |
| Heating plant annual maintenance cost per Mbtu of capacity: | $0 | $0 |
| 5. Reduction in heating maintenance: | $0 | $0 |
| 6. Other annual costs engendered by the lighting system: | | |
| Cost of spot replacement, one lamp (materials plus labor): | $31 | $55 |
| Lamp life (hours): | 12000 | 20000 |
| 7a. Cost of lamps annually (spot relamping): | $5,789 | $867 |
| 7b. or Annual group relamping cost: | | |
| Cost of spot replacement, one ballast (materials plus labor): | $25 | $50 |
| Ballast life (hours): | 85000 | 85000 |
| Lamps per ballast: | 2 | 1 |
| 8. Cost of ballast replacement | $335 | $186 |
| 9. Luminaire washing cost | $8 | $8 |
| 10. Annual insurance cost: | $0 | $0 |
| 11. Annual property tax | $0 | $0 |
| 12. Subtotal, annual power and maintenance (after tax): | $11,191 | $6,377 |
| Annual depreciation - | | |
| Accelerated (specify annual amount): | | |
| or straight line: | $0 | $1,800 |
| 13. Income tax effect of depreciation: | $0 | ($450) |
| | | |
| Annual energy and maintenance cost | $11,191 | $5,927 |
| | | |
| C. COMPARISONS | | |
| 1. Present worth | $75,094 | $57,768 |
| 2. Annual cost | $11,191 | $8,609 |

**FIGURE 6.3** Spreadsheet comparing fluorescent and HID industrial lighting systems.

changing lighting systems. This model evaluates the costs and savings achieved by converting to more energy-efficient lighting systems. An entire facility's new system can be compared with an existing system with a summary for the entire facility. An important feature of RENA is its ability to factor the savings in reduced air-conditioning costs during the cooling season that result from energy-efficient lighting systems. The increase in fuel costs during the heating season are also computed. The net savings gained annually and over the total rated life of the new lamps are shown.

The Lighting Life-Cycle Cost Analysis (LLC) model evaluates and compares an unlimited number of lamps to determine which lamp type or lighting system is the most cost-effective. The results show a true perspective of a lamp's cost over its life by evaluating lamp cost, energy cost, and labor cost and comparing the total cost to the amount of light produced. LCC shows how many lumen-hours (light over life) are produced for every penny spent. This model will also indicate the best lamp choice for competitive replacement.

The Return on Investment (ROI) model provides the user with an area-by-area, floor-by-floor, computer-generated analysis for numerous different lighted areas. The entire lighting operation can be seen at a glance, while detailing ROI for each area, lighting system, or entire facility. The ROI model evaluates all operating costs: lamps, energy, and labor. Total costs for all the various lamp types used throughout the facility are shown. Then, the use of value-added, energy-saving, cost-reducing lamps on proposed energy-efficient systems are shown. ROI also computes the environmental emissions avoided for carbon dioxide, sulfur dioxide, and nitrogen oxides.

The Economic Analysis (ECON) model determines which lighting system offers the actual lowest total cost. The costs of installing, maintaining, and operating different lighting systems are compared. ECON shows how many fixtures are required or what light levels (in footcandles) the system will maintain. The simple payback and life cycle cost information are provided to enable comparisons of proposed systems with an existing one.

The Group Relamp (GRP) model calculates the best relamping method and optimum relamping time by analyzing and comparing the cost and effects of three replacement methods: spot replacement, group replacement, and group-spot (group relamping with spot replacement of early burnouts). The model considers the annual lamp and labor costs for all three replacement methods and shows the number of years between relampings based on either a maintained light level (illuminance) or maximum savings.

Often auditors using lighting software will need to look up information on various lamps. Information on the most popular lamp types in the OSRAM SYLVANIA catalog is available to the SYLVIA user in the form of a lamp database that may be accessed at the touch of a button. When the user is working within a particular model, pressing the BROWSE button brings up a pop-up window that allows the user to scroll through the database. A SEARCH button brings up an expanded window that allows the user to enter criteria to initiate a search for a lamp. The chosen lamp will be automatically entered in the model. A shortcut allows lamps that have been previously used for a customer to be selected by using the ENTERED LAMPS button.

A luminaire database has been added to this version of SYLVIA PC. It is used to help the user select a coefficient of utilization (CU) needed in some of the models. About 35 sample fixtures with an average CU for each fixture are available. From the file menu, the luminaire database is accessed by selecting the LUMINAIRE DATABASE option. The worksheet button next to the CU field also accesses the database. The user may scroll through the description of fixtures or change views of the picture to select SKETCH, DISTRIBUTION, or SECTION. A box located next to the description requires the user to enter a value for the percent of ceiling reflectance. A wall reflectance grid appears and the user selects a coefficient by selecting a value where the wall reflectance and the room cavity ratio (RCR) meet. The resulting CU value will be entered in the model automatically.

An expanded context-based HELP system provides on-line help any time the user gets stuck or is not sure what to enter in a particular field. When the F1 key is pressed, pop-up windows provide information regarding the field the user is in. The F2 key accesses more detailed information. A HELP TIPS button turns on messages every time the mouse cursor is over a button.

## LightPAD

The Electric Power Research Institute (EPRI) has an audit program called LightPAD that is available through its 700 member utilities. LightPAD, a simple computer program designed for utility lighting auditors and audit contractors, is intended to greatly expedite the work of lighting auditors (see Table 6.3). Its name is derived from Lighting Portable Audit and Design tool. LightPAD is a flexible auditing tool, not an auditing procedure. It will accommodate a wide range of auditing styles and utility conservation programs. LightPAD can provide a quick estimate of the lighting energy use of a building, or it can be used as a detailed analysis tool for determining both lighting energy use and current light levels. Hourly lighting use patterns, differing energy rates, and other variables may be input so the LightPAD program can more accurately recommend upgrade alternatives that both reduce energy consumption and improve lighting quality. All data can be input directly into the LightPAD program on-site, the data can be verified on-site, and upgrade alternatives can be compared on-site.

Using this tool, auditors can describe the existing space and lighting system layouts and evaluate alternative or retrofit lighting systems. The alternative systems can be compared with the existing system on the basis of several measures—light levels, power requirements, cost of installation, and cost of system operation. Reports can be generated on any of these comparative measures.

LightPAD operates on either notebook or POA computers, so a notebook or laptop computer can be used on-site for data collection. There are several advantages to having an on-site tool for data collection. The first is that it speeds the data collection process. The second advantage is that results can be

**TABLE 6.3** LightPAD Program Overview

| What LightPAD Does | What LightPAD Doesn't Do |
|---|---|
| **Calculates:** | |
| Lighting levels (lumen method) | Point-by-point illuminance |
| Daylight analysis | Isolux maps of light levels |
| Energy use—cost and savings | Graphic descriptions |
| Demand—cost and savings | Complex financial analysis (export to COMTECH if further financial analysis is required) |
| Maintenance costs | |
| Simple payback | |
| **Compares:** | |
| Estimated to measured data | |
| Maintenance factors | |
| Diversity and lighting use factors | |
| Retrofit alternatives | |
| **Defines:** | |
| Existing space | |
| Existing lighting components | |
| **Reports (writes to files):** | |
| Project information | |
| Lighting system summaries | |
| Retrofit schedules | |
| Materials lists | |
| Space-by-space results | |
| **Tracks:** | |
| Component costs | |
| Rebates | |
| Retrofit costs | |
| Project status | |
| **Uses:** | |
| Utility-specific rebates | |
| Component databases | |
| Lighting use schedules | |
| Electric rate schedules | |
| Space type characteristics | |

checked before the audit team leaves the site, so auditors should not have to return at a later date to recheck questionable data. Also, the data can be checked within the program at several levels—the building level, the space level, and the component level.

The program can also be used to automatically create four files, representing the base (existing) condition and three alternatives. These files contain the total building kW hourly profiles for the lighting alternatives developed in LightPAD The files can then be exported to another EPRI program (COMTECH) to perform more complex financial analysis and to integrate the lighting portion into a total building audit.

## Program Inputs

*Space definition:* Room dimensions, surface reflectances, maintenance factors, ceiling plane, lighting plane, work plane, occupancy schedules, and relamping practices. The auditor can validate and adjust LightPAD calculations by measuring actual light levels and those factors that may modify light characteristics.

*Lighting equipment:* The program includes information on some general fixture and control types. Generic fixture data includes luminaire type, lamps, and ballasts. Generic control data includes control type. The auditor may also enter actual luminaire data, based on manufacturer's information. IESNA format files can be entered automatically. Cost information must be provided by the user.

*Utility information:* LightPAD calculates simple kilowatts (kW) and kilowatt-hours (kWh), based on estimated occupancy. Utility rates and rebate information can also be entered.

## Reports

LightPAD performs a number of lighting calculations and prepares a report based on the inputs described. The report includes:

1. **Space light level analysis:** Illuminance in footcandles calculated by the zonal-cavity method using luminaire photometric data
2. **Power use analysis:** kW, kWh, and a lighting power density in watts per square foot
3. **Financial analysis:** Capital and operating cost, net cost after rebate, and simple payback
4. **Material summary:** Type, quantity, and cost of all equipment used in the retrofit alternatives

LightPAD 2.0, an upgrade for the original 1.0 version, was released in 1996. The upgrade further simplifies the lighting auditing process. The range of capabilities virtually automates a lighting audit. The enhanced computer program employs the more intuitive buttons, layouts, and tabs available in the Windows platform. Version 2.0 is designed to run on a PC with Windows 3.1 for mouse and keyboard entry or Windows for Pen Computing 1.0 that supports "pen" and keyboard entry. Use of a 486 (or better) processor and a color monitor will enhance performance.

# InSite

XENERGY, a company with expertise in management, engineering, and economic consulting, has a commercial energy marketing and implementation system available called InSite. InSite is an easy-to-use field tool that allows utility customer representatives to collect data, analyze energy use, and recommend promotional and conservation measures for commercial and industrial customers. The computer program is designed for quick and easy data collection and powerful measure analysis. Created from the extensive field experience of XENERGY engineers, this Windows-based program helps users learn and use it efficiently.

Lighting is just one of the analysis modules contained in the InSite program. Other modules are: HVAC Equipment and Shell; Motors; Refrigeration; Water Heating and Other Equipment/Process. Cost and analysis inputs are restored in libraries of reference information that the user can edit and maintain. Promoted to turn data such as energy-use patterns and equipment profiles into useful, strategic information, the program should help utilities identify and serve customers who may benefit from their company's incentive programs.

## AUDIT 2000

Philips Lighting announced AUDIT 2000 software at the annual conference of the IESNA in August 1995. A Philips representative responded to inquiries in June 1996 that "Philips is not selling this program, and it is not available to the public." The author has concluded that either the announcement for availability was premature, or this software is only intended for and available to Philips personnel.

## Green Lights Programs

### Green Lights Decision Support System (GL/DSS)

The Green Lights Program is an innovative, voluntary program sponsored by the U.S. Environmental Protection Agency (EPA) that encourages major U.S. corporations and other organizations to install energy-efficient lighting technologies. Organizations that become Partners and make a commitment through a Memorandum of Understanding (MOU) will profit by (1) lowering their electric bills, (2) improving lighting quality, and (3) increasing worker productivity. The environment benefits by a reduction in air pollution caused by electric generation.

Part of the Green Lights Program support, the GL/DSS is a computer tool that provides help in identifying applicable lighting upgrade (retrofit) technologies and quantifying costs and benefits. The GL/DSS software package is designed to meet the following objectives:

1. To organize and reduce the effort involved in collecting building data
2. To provide room-by-room lighting upgrade recommendations that maximize pollution prevention and profitability (as defined by the GL MOU)
3. To improve the ability of GL Partners to work with consultants, designers, lighting management companies, and vendors to achieve upgrades that meet Green Lights program goals.

Based on lighting survey data, the GL/DSS evaluates existing lighting conditions and selects appropriate upgrade technologies from large databases that include performance and cost data. The lighting upgrade packages recommended by the software will maximize energy savings and maintain or improve lighting quality.

Users may identify preferences for specific lighting technologies or approaches that either should or should not be applied to a building or designated areas of the building. Preferences may be specified for:

1. luminaire replacement versus upgrade,
2. exclusion of specific retrofit (upgrade) technologies,
3. lighting maintenance method,
4. task lighting strategy, and
5. target illuminance (footcandles, FC) definition.

The GL/DSS allows Partners to make informed upgrade (retrofit) decisions and facilitates project planning by producing reports. The two reports of greatest value to the lighting auditor are the Building Survey Report that provides a comprehensive listing of all information collected during the lighting survey data collection process and the Room Definition and Analysis Summary Report that describes for each room in the building such things as quantity, wattage, and layout of existing and retrofit luminaires, illuminance (footcandles) levels, analysis method used, and unit power density.

### Quikalc

Another Green Lights computer program, Quikalc, is designed to accurately estimate the energy, environmental, and financial results of a specific lighting upgrade. The program compares the existing and upgrade option performance of a lighting system. Quikalc can calculate the performance of only one luminaire type at a time; therefore, it is a burdensome task to use it to analyze the lighting in a building that contains several luminaire types, because such analysis requires several runs of the program. Cash flows must then be manually aggregated into a combined project cash flow. Existing luminaire, retrofit, and operating data are entered on up to 24 input screens. The results may be immediately displayed in a "results window" that appears at the bottom of the screen.

While Quikalc calculates the performance of a retrofit option selected by the user, it does not recommend any specific upgrades. Therefore, knowledge of lighting quality and compatibility of equipment are essential for selecting efficient, high-quality retrofit options when using this program.

### A Comparison of Quikalc and GL/DSS

Both Quikalc and GL/DSS provide:

1. Assistance in identifying upgrade options that maximize energy savings and in calculating the internal rate of return
2. Calculation of annualized maintenance costs and light output based on group or spot relamping
3. Calculation of energy and pollution reductions (in kW, kWh, dollars, carbon dioxide, sulfur dioxide, and nitrous oxides)
4. Calculation of benefits from the use of automatic controls, such as occupancy sensors and time-based switching
5. Option of inputting rebate information
6. A choice of either new luminaires or a retrofit of existing luminaires
7. Calculation of Internal Rate of Return (IRR) and Net Present Value (NPV) based on cash flows that may incorporate user-supplied inflation factors

Note that neither program will analyze daylighting or dimming applications.

Important differences between Quikalc and GL/DSS include:

1. The analysis term, in years, can be modified in Quikalc but not in GL/DSS, which has the term fixed at 20 years.
2. Quikalc calculates the relative light output of an upgrade based on lamp lumens, ballast factor, luminaire efficiency, and depreciation effects, while GL/DSS calculates the average maintained horizontal illuminance (FC) based on an internal database of luminaire photometrics.
3. Quikalc allows users to input costs for individual components, while GL/DSS provides a single lump sum material cost for the upgrade system that can be modified only as a lump sum.
4. Quikalc permits users to determine an approximate savings from air-conditioning operating costs, while GL/DSS conservatively ignores air-conditioning savings.

The prime advantages of Quikalc are:

1. It provides a one-to-one comparison of an existing lighting system and an upgrade option.
2. It is quick and easy to use for screening or selecting lighting retrofit options.
3. It is excellent for performing sensitivity ("What if?") analyses by changing the values of key variables.
4. It provides a thorough, pretax life-cycle cost analysis of a given lighting retrofit project.

The important limitations of Quikalc are:

1. It does not suggest upgrade options.
2. It does not compile a buildingwide survey (it analyzes only one luminaire type at a time).
3. It does not directly analyze task/ambient systems.

The prime advantages of GL/DSS are:

1. It recommends upgrade that maximize energy savings when inputs regarding tasks and target light levels are provided.
2. It provides a comprehensive after-tax financial analysis.
3. It educates users regarding energy-efficient technologies.
4. It facilitates the buildingwide survey effort required by the GL MOU.

The important limitations of GL/DSS are:

1. It requires additional effort during the survey (but achieves labor savings during the analysis step).
2. Output cannot be edited, and specific lighting upgrade solutions cannot be selected and analyzed.
3. It requires 10 megabytes of hard disk space.

## Calcu-Lite

Calcu-Lite, a simple lighting calculations computer program, can be used to approximate light loss (maintenance) factors and coefficients of utilization (CUs). The program usually is used in a more straightforward manner to calculate, using the IESNA zonal-cavity method (see Chapter 8, Fact Sheet 12), the number of fixtures required or the illuminance (light level) resulting when the light loss factor (LLF) and CU are known. Through a series of repeated iterations, the program can be used to zero in on a combination of LLF and CU that may be appropriate for the area being audited.

To demonstrate how the zonal-cavity method can be used to solve for CU, we will use the formula given in Chapter 8, Fact Sheet 12, and rearrange the terms to solve for CU.

$$CU = \frac{L \times W \times FC\ (\text{maintained})}{\#\ \text{fixtures} \times \#\ \text{lamps/fix} \times \text{lumens/lamp} \times LLF}$$

The user enters dimensions in each box—length (L), width (W), fixture to work plane (h)—and the program calculates the RCR (room cavity ratio). Next, the user enters a value for the CU, which is normally found in luminaire catalogs (however, for this calculation it is estimated). Then, data regarding the number of lamps per fixture, the catalog lumen rating for the lamps, and an estimated LLF are entered. Calcu-Lite calculates the average illuminance in footcandles (or lux) when the user prescribes the number of fixtures. The resulting light level value is compared to the field-measured value. If it is too high, the CU value is reduced. LLF may also need to be examined and reduced. The process continues until the desired value of illuminance is obtained.

The iterative process is best understood by using a case history example. A corporate dining room has a luminous ceiling consisting of 4-foot, 2-lamp strip lights equipped with 34-watt reduced-wattage fluorescent lamps. There are 120 fixtures in an area of 30 feet by 40 feet (1200 square feet). The power density is approximately 7–8 watts per square foot. The shielding media covering the entire ceiling is an "egg crate" lattice-work metal louver with openings of approximately 1 inch square. This metal louver blocks a lot of the light from the strip fixtures, mounted about 2 feet above the louver. In addition, there are about 15% burnouts and the lamps appear old. There is dirt on the sides of the louver openings. An average of 50 FC is measured at the dining room tables, located 2.5 feet from the floor.

Because the lamps are observed to be old and there is a high potential for both "energy-saving" (reduced-wattage) lamps and "energy-saving" ballasts, a lower ballast factor must be considered. Due to these factors, the LLF is estimated at 0.60 for the first iteration. It is estimated that about 75% of the light is being blocked, so the CU is estimated at 0.25. Figure 6.4 shows the Calcu-Lite screen calculating a light level of 84 FC—too high. On the second iteration, it is recalled that dirt on the louvers was not factored into the LLF, so it is reduced to 0.55, and low wall reflectances were not accounted for, so the CU is reduced to 0.20 (Figure 6.5). With these values, Calcu-Lite calculates a light level of 61.6 FC—still too high. On the third iteration (Figure 6.6), with LLF left at 0.55 and the CU estimated at 0.16, the light level calculates to 49.3 FC—close enough. So, this luminous ceiling has only 16% of the light transmitted through it initially and has lost 45% due to dirt, lumen depreciation, and ballast factor. The program proved valuable in calculating the CU more accurately than initial estimates that placed the CU at 0.25—50% too high!

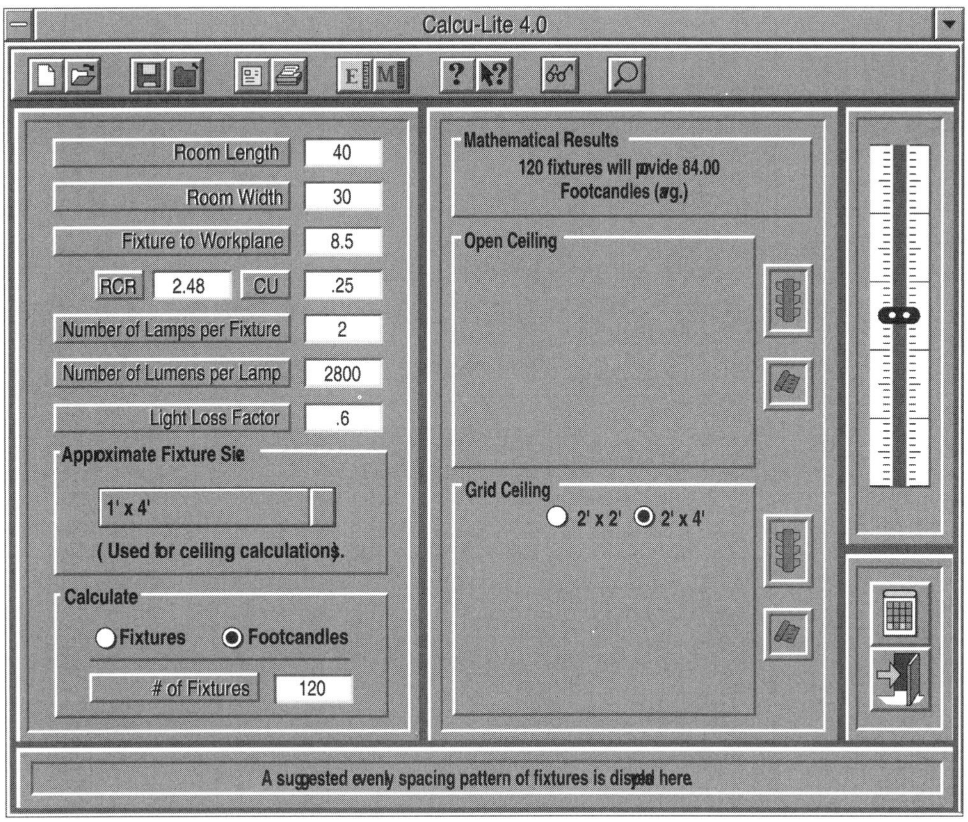


**FIGURE 6.4** Calcu-Lite screen showing light level calculations.

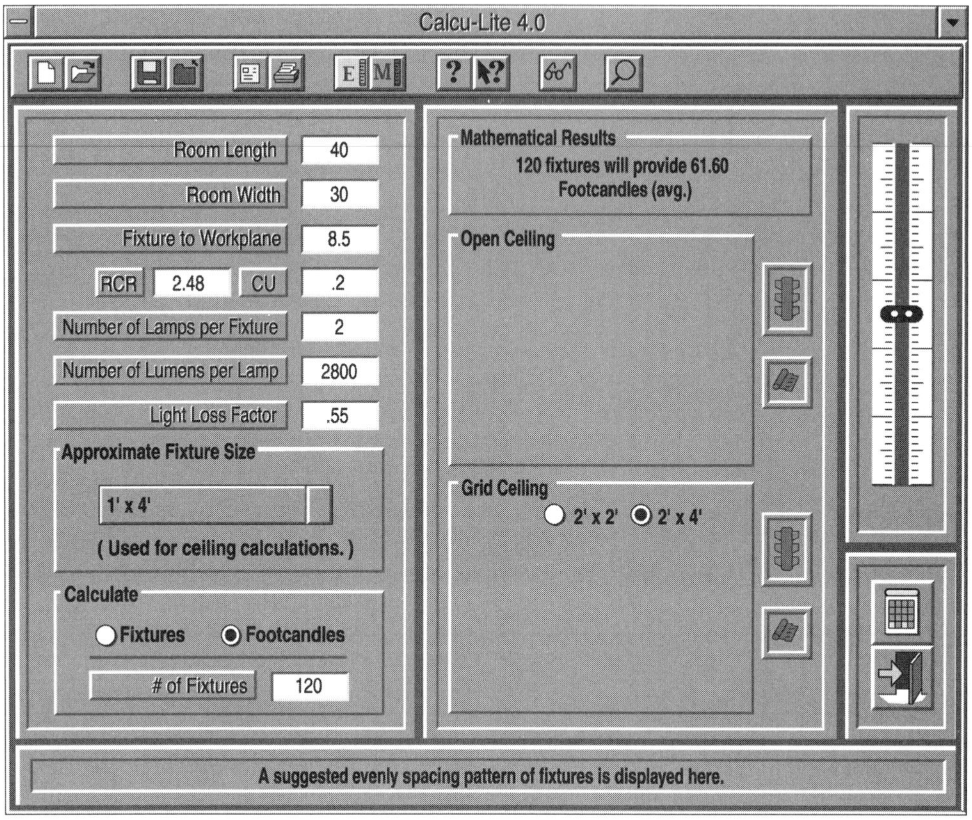


**FIGURE 6.5** Calcu-Lite screen showing revised light level calculations.

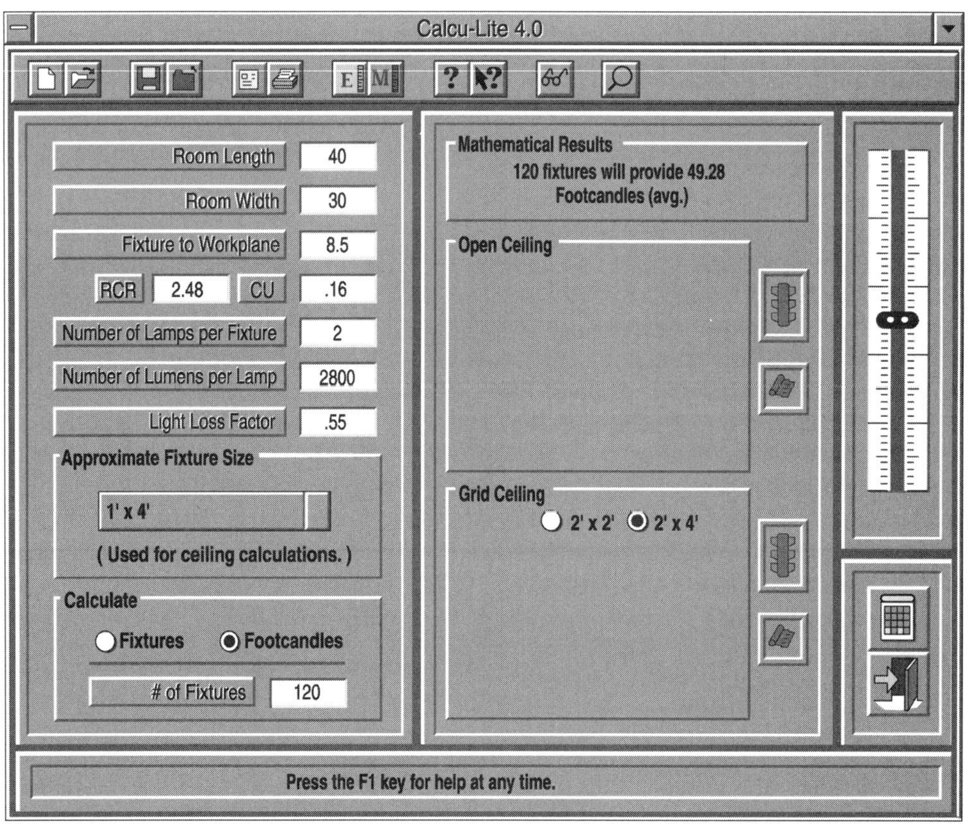


**FIGURE 6.6** Calcu-Lite screen showing third and final revision of light level calculations.

## DataMasters Lighting Audits Made Plain and Simple (L.A.M.P.S.)

This program is part of an auditing system that employs a hand-held Fujitsu Stylistic computer for performing lighting audits and surveys. The L.A.M.P.S. program is specifically tailored for retrofit applications and is particularly well suited for large companies that require standardized documentation among many different entities and are seeking a user friendly interface that will allow them to conduct lighting audits and surveys with speed, uniformity, and accuracy.

L.A.M.P.S. was developed from the perspective of a lighting contractor and is intended for use by anyone involved in the performance of lighting retrofit-related activities. The pen-based, hand-held computer permits quick and efficient data entry and generates documents for auditing, design, specifications, costing, analysis, sales presentation, materials ordering, and job tracking. Using this program virtually eliminates the normally lengthy re-entry and analysis of field audit data. The program capitalizes on a highly sophisticated, fully relational database architecture and provides a wide range of standard reports that can easily be customized to suit the user's needs. The best part is that all the reports are available immediately after the survey or audit is completed, before the audit team even leaves the site!

## Summary

There are many software programs available to aid the auditor in data collection and analysis. Software programs are tools, just as light meters and loggers are tools, and they must be used properly to provide meaningful results. No spreadsheet or computer program can supply the auditor with common sense or judgment. The auditor needs to look at computer-generated results with a skeptical eye, because there is something very convincing about computer generated numbers. At several points the auditor should ask, "Does this result make sense?" These sanity checks will help keep the audit results credible and the auditor in business.

# Chapter 7
# Sample Audit Reports

Reports are the final result of surveys and audits. They are the important record of the survey or audit findings and include the data collected and recommendations for upgrading the lighting systems of a facility. In our reports, we communicate to the client not only those recommendations that we have found meet the upgrade criteria but also those options that we investigated that we did not recommend. Unless this information is provided, those who read the report will always wonder "what if?" regarding those options not discussed in the report. If an option is not recommended due to long payback, it can be set aside until electric rates are high enough or payback criteria change, making the option more attractive.

## Sample Walk-Thru Survey Report

A typical walk-thru survey report (in this case performed for a nursing home) is shown in Figure 7.1. The report does not record a count of fixtures beyond a count made by the nursing home staff for the first few fixture types. The first line of the spreadsheet that summarizes the existing fixtures and the recommended changes reports a safety condition that is a potential liability situation. Although the simple payback (SPB) is 47 years, this recommendation has a priority 1, due to the safety and liability concerns. The recommended new fixture and lamp combination will provide an approximately 50% higher light level at no additional power costs.

The report is written in concise language that will help the nontechnical nursing home personnel to purchase and install the recommended equipment. Specific equipment recommendations are made that will meet the performance criteria. No alternatives are provided, and the end-user is responsible for the outcome resulting from substitution of alternative lighting equipment.

---

### A Report for Sunny Dale Lutheran Home

Effective Lighting Solutions, Inc., a lighting efficiency consulting firm, was contracted by Sunny Dale to provide recommendations for cost-effective lighting efficiency solutions.

We recommend that Sunny Dale Lutheran Home make the changes recommended in the accompanying spreadsheet that are identified as priority 1 or 2, for paybacks of 2.5 years or less.

**Line 1: Priority 1, Fixture Type 5**
On our visit January 23, 1996, we counted 266 of these round, surface-mounted circleline fixtures. While there is not a lot of savings for this line item, you will boost light levels by approximately 50%. Since the area is too dark now, you will want to do this retrofit for safety and liability reasons.

**Line 2: Priority 2, Fixture Type 1**
We counted about 96 of these round, surface-mounted, double circleline fixtures, which we recommend to also be replaced with the same ALTA fixture as used on Line 1. This brings the total count for the SMA-38 fixture to 362. A verification count should be made before an order is placed.

*Note: No fixture count was made for the recommendations that follow.*

**Line 3: Priority 1, Fixture Type 8**
The 75R30 lamp is no longer being manufactured. Replace this bulb throughout your facility with the OSRAM SYLVANIA 50PAR30LN/CAP/WFL. Catalog sheet included.

**Lines 14, 4, 15, 5, 10: Priority 1 and 2, Fixture Type 2, 3, 13, 14, 16**
The F40T12CW lamp you are currently using is no longer being manufactured. We recommend that you convert to the newer-technology 1-inch T8 lamps. We recommend the OSRAM SYLVANIA Octron FO32/735 for a somewhat warmer appearance than your cool-white lamps now provide. If you want to stay with the cool color, use the FO32/741. Catalog sheet included.

Sunny Dale Nursing Home
Southern, CA

Lighting Survey
Performed April 19, 1995 & 1/23/96
by Effective Lighting Solutions, Inc.
Columbus, Ohio 43205
Survey conducted by John Fetters
Certified Lighting Efficiency Professional

| Line | Priority | Fix Type | Existing Light Fixture Description | Est Watts | Recommended Change | Watts | Power Reduction | Est hrs per year | Annual kWh Savings | Estimated Annual Savings/Fix | Estimated Material Cost | Simple Payback (yrs) | Comments |
|---|---|---|---|---|---|---|---|---|---|---|---|---|---|
| 1 | 1 | 5 | Round, surface mount, w/FC12T9 circleline lp | 40 | ALTA SMA-38 w/2-D lamp & ELB | 38 | 2 | 8,760 | 18 | $1.58 | $75.00 | 47.6 | Area too dark now; est 50% more light - DO for Safety! |
| 2 | 2 | 1 | Round, surface mount, w/32/22 circleline lps | 70 | ALTA SMA-38 w/2-D lamp & ELB | 38 | 32 | 8,760 | 280 | $25.23 | $75.00 | 3.0 | Equivalent light |
| 3 | 1 | 8 | Various fixtures, w/75R30 lamps | 75 | 50w T-H PAR30 flood (50PAR30LN/CAP/WFL) | 50 | 25 | 8,760 | 219 | $19.71 | $7.00 | 0.4 | Low cost, short payback, high priority (EPAct) |
| 4 | 1 | 3 | 4', 4-lamp, w/F40CW lamps | 190 | 4-T-8 lamps + electronic ballast (ELB) | 106 | 84 | 8,760 | 736 | $66.23 | $30.00 | 0.5 | Long burn hours = short payback = high priority (EPAct) |
| 5 | 1 | 14 | Cove lights, 2-lamp, w/F40CW lamps | 95 | 2-T-8 lamps + electronic ballast (ELB) | 60 | 35 | 8,760 | 307 | $27.59 | $23.00 | 0.8 | Long burn hours = short payback = high priority (EPAct) |
| 6 | 1 | 18 | 4-chandeliers w/4-G lamps (Ruud Comm Ctr) | 1,600 | 15 watt Compact fluorescent G40 lamp | 240 | 1,360 | 3,000 | 4,080 | $367.20 | $320.00 | 0.9 | Medium cost, less than 1 yr payback |
| 7 | 1 | 9 | Various fixtures w/75 w, incand A lamp | 75 | SLS-20 integral CFL | 20 | 55 | 4,000 | 220 | $19.80 | $20.00 | 1.0 | Low cost, short payback, high priority |
| 8 | 1 | 7 | Surface mtd socket, w/G40 incand lamp | 100 | 15 watt Compact fluorescent G40 lamp | 15 | 85 | 2,200 | 187 | $16.83 | $20.00 | 1.2 | Low cost, short payback, high priority |
| 9 | 1 | 10 | Various fixtures w/60 w, incand A lamp | 60 | 42 w capsulite, tungsten-halogen lamp | 20 | 40 | 4,000 | 160 | $14.40 | $20.00 | 1.4 | Low cost, short payback, high priority |
| 10 | 1 | 16 | 4x4, 6-lamp, w/F40CW lamps | 285 | 6-T-8 retrofit lamps + 1-2 lp & 1-4 lp ELB | 180 | 105 | 4,000 | 420 | $37.80 | $55.00 | 1.5 | Medium cost, less than 2 yr payback (EPAct) |
| 11 | 1 | 12 | Beauty shop w/21-G lamps beside mirrors | 2,100 | 15 watt Compact fluorescent G40 lamp | 315 | 1,785 | 1,560 | 2,785 | $250.61 | $420.00 | 1.7 | Complaint of "too hot" from incand G-lamps |
| 12 | 2 | 15 | 2x2, 2-lamp, w/FB40CW U lamps | 95 | 2-T-8 retrofit lamps + ELB | 60 | 35 | 4,000 | 140 | $12.60 | $29.00 | 2.3 | EPAct affected lamp - October 31,95 deadline |
| 13 | 2 | 19 | Exit light w/2-20T6-1/2 incand lamps | 40 | New Conservation Alliance LED exit lights | 2 | 38 | 8,760 | 333 | $29.96 | $68.00 | 2.3 | Energy savings and maintenance savings |
| 14 | 2 | 2 | 4', 2-lamp, w/F40CW lamps - std ballast | 95 | 2-T-8 lamps + electronic ballast (ELB) | 60 | 35 | 3,000 | 105 | $9.45 | $23.00 | 2.4 | EPAct affected lamp - October 31,95 deadline |
| 15 | 2 | 13 | 2x4, 2-lamp, w/F40CW lamps | 95 | 2-T-8 lamps + electronic ballast (ELB) | 60 | 35 | 3,000 | 105 | $9.45 | $23.00 | 2.4 | EPAct affected lamp - October 31,95 deadline |
| 16 | 3 | 4 | 8', 2-lamp, w/F96WM lamps | 140 | 4-T-8 lamps + ELB + Harris conversion kit | 106 | 34 | 8,760 | 298 | $26.81 | $90.00 | 3.4 | High cost, small savings = low priority |
| 17 | 3 | 11 | Residence baths w/3-G lamps over sinks | 180 | 11 watt Compact fluorescent G30 lamp | 33 | 147 | 550 | 81 | $7.28 | $60.00 | 8.2 | Short burn hours = long payback = low priority |
| 18 | 3 | 17 | Surface mtd, w/FC6T9 circleline fluorescent | 28 | New fixture w/13 w TT CFL lp (damp rated) | 17 | 11 | 4,380 | 48 | $4.34 | $35.00 | 8.1 | High cost, long payback, do after others done |
| 19 | 3 | 20 | Exit light w/2-F6T5 fluor lamps | 20 | New LED exit lights | 3 | 17 | 8,760 | 149 | $13.40 | $70.00 | 5.2 | Phase in as units are replaced to lower maint costs |
| 20 | 3 | 21 | Exit light w/AC/DC lamps | 36 | New LED exit lights | 3 | 33 | 8,760 | 289 | $26.02 | $130.00 | 5.0 | Phase in as units are replaced to lower maint costs |
| | | 6 | Recessed, square, w/FC8T9 circleline lamp | 30 | None at this time | 30 | 0 | | 0 | $0.00 | $0.00 | | |

**FIGURE 7.1** Sample walk-thru survey spreadsheet report.

When you make this conversion, you will need to change the ballast. We recommend a high-frequency electronic ballast, which will operate the T8 lamps more efficiently and save on your electric bill. You have both 2-lamp and 4-lamp fixtures (plus one 6-lamp fixture) that will need either the 2-lamp ballast or the 4-lamp ballast. (Note that the fixtures that use two 2-lamp ballasts can be rewired for one 4-lamp ballast for a good savings.) We recommend:

For the 2-lamps, use OSRAM SYLVANIA QT2X32/120LP.
For the 4-lamps, use OSRAM SYLVANIA QT4X32/120LP.

Catalog sheet included.

**Lines 6, 8, 11: Priority 1, Fixture Type 18, 7, 10**
Several locations now use the G (globe) incandescent lamp. By replacing these incandescent lamps with compact fluorescents, you will save energy and electric costs, cut down on heat (especially in the beauty shop), and reduce maintenance costs (globes have a 10,000-hour rated life). We recommend these compact fluorescent replacements:

For the G30 bulbs, use OSRAM SYLVANIA CF11EL/G30/827/MED.
For the G40 bulbs, use OSRAM SYLVANIA CF15EL/G40/827/MED.

Catalog sheet included.

**Line 7: Priority 1, Fixture Type 9**
Various incandescent fixtures use a 75-watt A bulb. We recommend that you change the bulb in each of these fixtures to an integral-ballasted (one-piece) compact fluorescent unit such as Philips SLS-20.

**Line 9: Priority 1, Fixture Type 10**
Various incandescent fixtures use a 60-watt A bulb. We recommend that you change the bulb in each of these fixtures to a 42-watt tungsten-halogen bulb with reduced wattage but equivalent light. We recommend OSRAM SYLVANIA 42MB/CAP, which is available in both 120-volt and 130-volt versions. Catalog sheet included.

**Line 12: Priority 2, Fixture Type 15**
There are a number of 2 × 2 fixtures that currently use the fluorescent 6-inch leg spacing U-tube. You can convert these to a T8 retrofit lamp, OSRAM SYLVANIA catalog number FBO32/835/6 for neutral color or FBO32/841/6 for cool color. Use the 2-lamp electronic ballast, OSRAM SYLVANIA catalog number QT2X32/120LP.

**Line 13: Priority 2, Fixture Type 19**
We recommend that the EXIT light units that presently use two 20-watt T6-1/2 incandescent lamps be replaced with new light-emitting diode (LED) units. This upgrade will save energy, but it will especially save in maintenance costs because the LED is a solid-state device with a rated life over 25 years! We recommend the Eagle II from Conservation Alliance. Catalog sheets included.

Your investments in lighting efficiency will pay off in the near term, plus you will be helping to care for our earth by reducing the air pollution caused from electrical generation.

I have enjoyed helping you and expect that you will call me if I can be of further assistance.

**John L. Fetters, Certified Lighting Efficiency Professional**

---

## Intermediate Audit Reports

Two summary spreadsheets for two different intermediate audits are shown in Figures 7.2 and 7.3. Figure 7.2 is part of an intermediate audit performed for a manufacturing plant in Ohio. This audit was done for a performance contractor, who wanted to know only of those retrofit options that would save energy and have a payback of less than 4 years.

The first page of Figure 7.2 shows the data collected for the fixtures identified as potential retrofit candidates, along with the fixture counts, fixture

An Effective Lighting Solutions, Inc. Survey & Analysis for the ABC Company, Inc.

Factory B
Our Town, Oh
LIGHTING RETROFIT RECOMMENDATIONS

| LOCATION Building A | QTY Fix | RETROFIT DESCRIPTION | QTY LPS | Fixture WATTS | Estimated ANNUAL HRS USE | Estimated ANNUAL KWH | Estimated KWH SAVINGS | \$0.0476 Estimated RETROFIT SAVINGS | Estimated RETROFIT CO\$T | SPB Yrs | Qualify? |
|---|---|---|---|---|---|---|---|---|---|---|---|
| Towmotor Repair | 85 | Electronic interval timer control | 2 | 76 | 1,664 | 10,749 | 16,124 | $767 | $1,200 | 1.6 | Yes |
| Men's Restroom | 16 | Electronic interval (switchplate) tmr/contactor | 2 | 76 | 2,000 | 2,432 | 2,627 | $125 | $475 | 3.8 | Yes |
| | 6 | Electronic interval (switchplate) tmr/contactor | 2 | 76 | 2,000 | 912 | 985 | $47 | $475 | 10.1 | No |
| | 8 | Electronic interval (switchplate) tmr/contactor | 4 | 175 | 2,000 | 2,800 | 3,024 | $144 | $475 | 3.3 | Yes |
| Machine Shop Electric Staging | 18 | Disconnect power | 2 | 76 | 0 | 0 | 5,691 | $271 | $70 | 0.3 | Yes |
| | 10 | Disconnect power | 1 | 40 | 0 | 0 | 1,664 | $79 | $70 | 0.9 | Yes |
| | 4 | Disconnect power | 2 | 180 | 0 | 0 | 2,995 | $142 | $70 | 0.5 | Yes |
| Vertical Pump Rm | 10 | Electronic interval timer switch/contactor | 1 | 40 | 2,000 | 800 | 864 | $41 | $575 | 14.0 | No |
| | 10 | Electronic interval timer switch/contactor | 2 | 76 | 2,000 | 1,520 | 1,642 | $78 | $575 | 7.4 | No |
| Shipping Dock Offic | 7 | Electronic interval (switchplate) tmr/contactor | 4 | 175 | 2,000 | 2,450 | 8,281 | $394 | $475 | 1.2 | Yes |
| Building B Offices above docks | 6 | Electronic interval (switchplate) tmr/contactor | 4 | 175 | 2,000 | 2,100 | 2,268 | $108 | $475 | 4.4 | No |
| Oven area | 3 | Convert to T8 + 4-lp ELB | 4 | 121 | 4,160 | 1,510 | 674 | $32 | $180 | 5.6 | No |
| Break room | 8 | Occupancy sensor | 4 | 175 | 1,500 | 2,100 | 3,724 | $177 | $425 | 2.4 | Yes |
| Hi speed pump rm | 16 | Rewire 2 fixtures for safety when remainder controlled by occupancy sensor | 2 | 76 | 2,000 | 3,764 | 8,220 | $391 | $965 | 2.5 | Yes |

**FIGURE 7.2A** Sample intermediate audit spreadsheet. Page 1—Data collection.

An Effective Lighting Solutions, Inc. Survey & Analysis for the ABC Company, Inc.

Factory B
Our Town, Oh
1995 LIGHTING SURVEY

| LOCATION | QTY Fix | FIXTURE DESCRIP | LAMP DESCRIP | LAMP WATTS | LPS/ FIX | BALLAST TYPE | LPS/ BAL | FIXTURE WATTS | TOTAL WATTS | Estimated ANNUAL HRS USE | Estimated ANNUAL KWH |
|---|---|---|---|---|---|---|---|---|---|---|---|
| Building A<br>Towmotor Repair<br>South Warehouse | 85 | 4 ft 2-lp open industr fluor | F40T12WM | 34 | 2 | Mix | 2 | 76 | 6,460 | 4,160 | 26,874 |
| Men's Restroom | 16 | 1X4 2-lp fluorescent | F40T12WM | 34 | 2 | Mix | 2 | 76 | 1,216 | 4,160 | 5,059 |
| | 6 | 2x4 2-lp fluorescent | F40T12WM | 34 | 2 | Mix | 2 | 76 | 456 | 4,160 | 1,897 |
| Quality Lab | 8 | 2x4 4-lp fluorescent | F40T12WM | 34 | 4 | Mix | 2 | 175 | 1,400 | 4,160 | 5,824 |
| Machine Shop<br>Electric Staging | 18 | 4 ft 2-lp open strip fluor | F40T12WM | 34 | 2 | Mix | 2 | 76 | 1,368 | 4,160 | 5,691 |
| | 10 | 4 ft 1-lp open strip fluor | F40T12WM | 34 | 1 | Mix | 1 | 40 | 400 | 4,160 | 1,664 |
| | 4 | 8 ft 2-lp open strip fluor | F96T12HOWM | 60 | 2 | Old | 2 | 180 | 720 | 4,160 | 2,995 |
| Vertical Pump Rm | 10 | 4 ft 1-lp open strip fluor | F40T12WM | 34 | 1 | Mix | 1 | 40 | 400 | 4,160 | 1,664 |
| | 10 | 4 ft 2-lp open strip fluor | F40T12WM | 34 | 2 | Mix | 2 | 76 | 760 | 4,160 | 3,162 |
| Shipping Dock Offic | 7 | 2x4 4-lp fluorescent | F40T12WM | 34 | 4 | Mix | 2 | 175 | 1,225 | 8,760 | 10,731 |
| Building B<br>Offices above docks | 6 | 2x4 4-lp fluorescent | F40T12WM | 34 | 4 | Mix | 2 | 175 | 1,050 | 4,160 | 4,368 |
| Oven area | 3 | 8 ft 4-lp open strip fluor | F40T12WM | 34 | 4 | Mix | 2 | 175 | 525 | 4,160 | 2,184 |
| Break room | 8 | 2x4 4-lp fluorescent | F40T12WM | 34 | 4 | Mix | 2 | 175 | 1,400 | 4,160 | 5,824 |
| Hi speed pump rm | 18 | 4 ft 2-lp open strip fluor | F40T12WM | 34 | 2 | Mix | 2 | 76 | 1,368 | 8,760 | 11,984 |
| Totals | 209 | | | | | | | Watts =<br>kW = | 18,748<br>18.7 | kWh = | 89,919 |

**FIGURE 7.2B** Sample intermediate audit spreadsheet. Page 2—Retrofit recommendations.

YOUR YMCA
Lighting Survey

| Location | Qty Fix | Fixture Descrip. | Qty Lps | WATTS | ANNUAL HRS USE | ANNUAL KWH | Incand Qty | Fluor Qty | Exit Qty | HID Qty | Incand KW | Fluor KW | Exit KW | HID KW | Incand KWH | Fluor KWH | Exit KWH | HID KWH |
|---|---|---|---|---|---|---|---|---|---|---|---|---|---|---|---|---|---|---|
| Exterior: | | | | | | | | | | | | | | | | | | |
| South side | 1 | "pointed" cobra-head | 1 | 455 | 4,380 | 1,993 | | | | 1 | | | | 0.455 | | | | 1,993 |
| South on bldg | 2 | Wallpack | 1 | 570 | 4,380 | 2,497 | | | | 2 | | | | 0.570 | | | | 2,497 |
| Service entrance | 1 | Wallpack | 1 | 100 | 4,380 | 438 | 1 | | | | 0.100 | | | | 438 | | | |
| North side | 2 | rooftop floods | 1 | 570 | 4,380 | 2,497 | | | | 2 | | | | 0.570 | | | | 2,497 |
| West side | 1 | Wallpack | 1 | 100 | 4,380 | 438 | 1 | | | | 0.100 | | | | 438 | | | |
| SW corner | 1 | Rental-pole mtd yard | 1 | 180 | 4,380 | 788 | | | | 1 | | | | 0.180 | | | | 788 |
| Driveway pole | 1 | Rental flood | 1 | 285 | 4,380 | 1,248 | | | | 1 | | | | 0.285 | | | | 1,248 |
| Interior Spaces: | | | | | | | | | | | | | | | | | | |
| 2nd Fl Pickering Rm | 8 | 4' SM Wrap-Arnd Len | 4 | 1,520 | 3,000 | 4,560 | | 8 | | | | 1.520 | | | | 4,560 | | |
| 2nd Fl Corridor | 3 | 4' SM Wrap-Arnd Len | 2 | 294 | 3,000 | 882 | | 3 | | | | 0.294 | | | | 882 | | |
| | 2 | 2' SM Wrap-Arnd Len | 2 | 110 | 3,000 | 330 | | 2 | | | | 0.110 | | | | | | |
| | 3 | Round Incand SM | 2 | 600 | 3,000 | 1,800 | 3 | | | | 0.600 | | | | 1,800 | | | |
| 2nd Fl Rm 201 | 4 | 8' D/I Pend Mtd | 4 | 760 | 3,000 | 2,280 | | 4 | | | | 0.760 | | | | 2,280 | | |
| 2nd Fl Rm 203 | 2 | 8' D/I Pend Mtd | 4 | 380 | 3,000 | 1,140 | | 2 | | | | 0.380 | | | | 1,140 | | |
| 2nd Fl Rm 204 | 2 | 8' D/I Pend Mtd | 4 | 380 | 3,000 | 1,140 | | 2 | | | | 0.380 | | | | 1,140 | | |
| Mens RR | 1 | Round Incand SM | 2 | 200 | 2,000 | 400 | 1 | | | | 0.200 | | | | 400 | | | |
| | 2 | Incand Sm | 1 | 200 | 2,000 | 400 | 2 | | | | 0.200 | | | | 400 | | | |
| Kitchenette "ICE" | 1 | Incand SM | 1 | 500 | 3,000 | 1,500 | 1 | | | | 0.500 | | | | 1,500 | | | |
| Janitor's Closet | 1 | Incand SM | 1 | 100 | 1,000 | 100 | 1 | | | | 0.100 | | | | 100 | | | |
| Womens RR | 2 | Incand SM | 2 | 400 | 2,000 | 800 | 2 | | | | 0.400 | | | | 800 | | | |
| | 1 | Round Incand SM | 2 | 200 | 2,000 | 400 | 1 | | | | 0.200 | | | | 400 | | | |
| Kitchen | 5 | Blue Incand SM | 1 | 2,500 | 2,000 | 5,000 | 5 | | | | 2.500 | | | | 5,000 | | | |
| Mech Rm(N of Kit) | 2 | Bare Lamp Porcelins | 1 | 200 | 1,000 | 200 | 2 | | | | 0.200 | | | | 200 | | | |
| Mech(N of Pick Rm) | 3 | Bare Lamp Porcelins | 1 | 300 | 1,000 | 300 | 3 | | | | 0.300 | | | | 300 | | | |
| East Stairs | 2 | Round Incand SM | 2 | 400 | 3,000 | 1,200 | 2 | | | | 0.400 | | | | 1,200 | | | |
| '1st Fl Hall | 2 | Sm Aper Incand DL | 1 | 200 | 3,000 | 600 | 2 | | | | 0.200 | | | | 600 | | | |
| Pool Observ Rm | 8 | 2x4 troffer | 4 | 1,520 | 2,500 | 3,800 | | 8 | | | | 1.520 | | | | 3,800 | | |
| Pool | 19 | Mercury Vapor SM | 1 | 8,550 | 3,500 | 29,925 | | | | 19 | | | | 8.550 | | | | 29,925 |
| Stairs to W Lock | 2 | Round Incand SM | 2 | 400 | 3,000 | 1,200 | 2 | | | | 0.400 | | | | 1,200 | | | |
| | 1 | Bare Lamp Porcelin | 1 | 100 | 3,000 | 300 | 1 | | | | 0.100 | | | | 300 | | | |
| Hall between RR's | 5 | Large Recess Incand | 1 | 1,500 | 3,500 | 5,250 | 5 | | | | 1.500 | | | | 5,250 | | | |
| Men's RR | 1 | Round Incand SM | 2 | 200 | 2,000 | 400 | 1 | | | | 0.200 | | | | 400 | | | |
| | 2 | Incand SM | 1 | 200 | 2,000 | 400 | 2 | | | | 0.200 | | | | 400 | | | |
| Reception Area | 6 | Large Recess Incand | 1 | 1,800 | 3,500 | 6,300 | 6 | | | | 1.800 | | | | 6,300 | | | |
| | 1 | Incand table lamp | 1 | 88 | 2,000 | 176 | 1 | | | | 0.088 | | | | 176 | | | |
| Counter Area | 2 | Recessed Incand DL | 1 | 200 | 3,500 | 700 | 2 | | | | 0.200 | | | | 700 | | | |
| Office bhnd counter | 6 | Recessed 2x4 troffers | 4 | 1,140 | 3,500 | 3,990 | | 6 | | | | 1.140 | | | | 3,990 | | |
| Director's Office | 2 | 8' D/I Pend Mtd | 4 | 390 | 3,500 | 1,365 | | 2 | | | | 0.390 | | | | 1,365 | | |
| Room 105 | 2 | 8' D/I Pend Mtd | 4 | 390 | 3,000 | 1,170 | | 2 | | | | 0.390 | | | | 1,170 | | |
| Closet | 1 | Incand glass SM | 2 | 200 | 1,000 | 200 | 1 | | | | 0.200 | | | | 200 | | | |
| Area into Gym | 8 | Large recess Incand | 1 | 2,400 | 3,000 | 7,200 | 8 | | | | 2.400 | | | | 7,200 | | | |
| Fulton Gym | 14 | Merc Vapor Industrial | 1 | 6,370 | 3,500 | 22,295 | | | | 14 | | | | 6.370 | | | | 22,295 |
| Stairs to Fitness Ctr | 2 | Round Incand SM | 2 | 400 | 3,000 | 1,200 | 2 | | | | 0.400 | | | | 1,200 | | | |
| | 1 | Bare Lamp Porcelin | 1 | 100 | 3,000 | 300 | 1 | | | | 0.100 | | | | 300 | | | |
| Corridor Cardio-fit | 3 | Large recess Incand | 1 | 900 | 3,000 | 2,700 | 3 | | | | 0.900 | | | | 2,700 | | | |
| Cardio-fitness Rm | 6 | Recessed 2x4 troffers | 4 | 1,140 | 3,000 | 3,420 | | 6 | | | | 1.140 | | | | 3,420 | | |
| | 2 | Recessed 2x4 troffers | 4 | 380 | 3,000 | 1,140 | | 2 | | | | 0.380 | | | | 1,140 | | |
| Stairs-Martin's Gym | 3 | Round Incand SM | 2 | 600 | 3,000 | 1,800 | 3 | | | | 0.600 | | | | 1,800 | | | |
| Martin's Gym (146) | 48 | 4 ' Industrial Fix | 2 | 4,560 | 3,000 | 13,680 | | 48 | | | | 4.560 | | | | 13,680 | | |
| Gym Office 1 | 1 | Recessed 2x4 troffers | 4 | 190 | 2,500 | 475 | | 1 | | | | 0.190 | | | | 475 | | |
| Gym Office 2 | 1 | Recessed 2x4 troffers | 4 | 190 | 2,500 | 475 | | 1 | | | | 0.190 | | | | 475 | | |
| Storage Rm | 2 | Bare Lamp Porcelin | 1 | 200 | 1,000 | 200 | 2 | | | | 0.200 | | | | 200 | | | |
| Men's RR | 2 | Round Incand SM | 2 | 400 | 3,000 | 1,200 | 2 | | | | 0.400 | | | | 1,200 | | | |
| Women's RR | 2 | Round Incand SM | 2 | 400 | 3,000 | 1,200 | 2 | | | | 0.400 | | | | 1,200 | | | |
| Short Corridor | 1 | 4' Wrap Around | 2 | 95 | 3,000 | 285 | | 1 | | | | 0.095 | | | | 285 | | |
| Long Corridor | 4 | 2x4 Troffers | 4 | 760 | 3,000 | 2,280 | | 4 | | | | 0.760 | | | | 2,280 | | |
| Men&Boy's Lock R | 8 | 2x4 Troffers | 3 | 1,120 | 3,000 | 3,360 | | 8 | | | | 1.120 | | | | 3,360 | | |
| Men's Health Club | 4 | 2x4 Troffers | 4 | 760 | 3,000 | 2,280 | | 4 | | | | 0.760 | | | | 2,280 | | |
| TV Area | 6 | 2x4 Troffers | 4 | 1,140 | 3,000 | 3,420 | | 6 | | | | 1.140 | | | | 3,420 | | |
| Locker Area | 4 | 2x4 Troffers | 4 | 760 | 3,000 | 2,280 | | 4 | | | | 0.760 | | | | 2,280 | | |
| Locker Area | 3 | 2x2 Troffers | 2 | 285 | 3,000 | 855 | | 3 | | | | 0.285 | | | | 855 | | |
| Showers | 4 | Jelly Jar Incand | 1 | 400 | 2,500 | 1,000 | 4 | | | | 0.400 | | | | 1,000 | | | |
| | 2 | 4' WL Fluor | 2 | 190 | 2,500 | 475 | | 2 | | | | 0.190 | | | | 475 | | |
| | 2 | 4' Fluor over Sinks | 1 | 100 | 2,500 | 250 | | 2 | | | | 0.100 | | | | 250 | | |
| Excersize Rm | 4 | 2x2 Troffers | 2 | 380 | 3,000 | 1,140 | | 4 | | | | 0.380 | | | | 1,140 | | |
| Weight Rm | 8 | 2x4 Troffers | 4 | 1,520 | 3,000 | 4,560 | | 8 | | | | 1.520 | | | | 4,560 | | |
| Stairs | 1 | 4' Wrap Around | 2 | 95 | 3,000 | 285 | | 1 | | | | 0.095 | | | | 285 | | |
| Corridor | 3 | 2x2 Troffers | 2 | 285 | 3,000 | 855 | | 3 | | | | 0.285 | | | | 855 | | |
| Weight Rm | 8 | 4' WL Fluor | 2 | 760 | 3,000 | 2,280 | | 8 | | | | 0.760 | | | | 2,280 | | |
| Child Care Rm | 8 | 2x4 Troffers | 4 | 1,520 | 3,000 | 4,560 | | 8 | | | | 1.520 | | | | 4,560 | | |
| Closet | 1 | Bare Lamp Porcelin | 1 | 100 | 1,000 | 100 | 1 | | | | 0.100 | | | | 100 | | | |
| Nautulus Rm | 6 | 2x4 Troffers | 4 | 1,140 | 3,000 | 3,420 | | 6 | | | | 1.140 | | | | 3,420 | | |
| Exit Lights: | 12 | Incandescent 2-lp | 2 | 600 | 8,760 | 5,256 | | | 12 | | | | 0.600 | | | | 5,256 | |
| | 297 | Totals | | 58,422 | | 184,333 | 76 | 169 | 12 | 40 | 16.588 | 24.254 | 0.600 | 16.980 | 45,402 | 72,102 | 5,256 | 61,243 |
| | | | Kw = | 58.42 | | | 25.6% | 56.9% | 4.0% | 13.5% | 28.4% | 41.5% | 1.0% | 29.1% | 24.6% | 39.1% | 2.9% | 33.2% |

**FIGURE 7.3** Sample intermediate audit data collection spreadsheet.

descriptions, lamp descriptions, lamp watts, lamps per fixture, ballast type, and lamps per ballast. From this data, the fixture watts value is determined and the total watts for each fixture type is calculated. From the estimated annual hours of use for each fixture type, the estimated annual kWh are calculated. The total number of fixtures is calculated, the total lighting power is totaled from the "Total Watts" (per fixture type) column, and the total annual lighting kWh is summed in the far righthand column.

The retrofit recommendations are shown in Figure 7.2B, with fixture types listed in the same order as the data collected and shown in Figure 7.2A. A retrofit description is provided for each fixture type listed, along with a calculation of the estimated annual kWh after retrofit, the estimated kWh savings (at an average electric cost of $0.0476/kWh), an estimated retrofit cost, and a calculation of the simple payback (SPB). The far righthand column indicates if the retrofit option qualifies, based on the payback criteria of less than 4 years.

The intermediate survey results shown in Figure 7.3 are for a YMCA. The format of this report is similar to that of the previous report, with the exception that the total lighting power (kW), and the percentage of each major lighting type (incandescent, fluorescent, HID, and EXIT lights) are shown for quantity of fixtures, kW, and kWh. From the summary spreadsheet, the fixture types can be compared as shown in Table 7.1 (note that the figures in Table 7.1 are slightly different from those in Figure 7.3 because of the "rounding errors" that commonly occur in spreadsheets).

This information can help clients understand which lighting systems are costing the most and where their lighting dollars are being spent. With the additional information in other parts of the report not shown, the recommended retrofits are shown along with how they will reduce the facility lighting energy costs.

## Comprehensive Audit Reports

A spreadsheet with a list of 4-lamp 2 × 4 recessed troffers in a 8676-square-foot office area of a 2-million-square-foot manufacturing facility is shown in Figure 7.4. This was a Green Lights audit, so some of the information is taken from the Green Lights *Lighting Upgrade Manual* and is designated in the next paragraph with a (GL).

Spaces are listed by name, function, or the name of the occupant. Data listed are the net area of each office, quantity of fixtures in each office or area, (GL) fixture type, fixture size, lamp type, a lamp description (RW means reduced wattage; STD means standard), lamp watts, lamps per fixture, (GL) ballast type, lamps per ballast, and (GL) fixture watts. The "Total Watts" value is calculated by the spreadsheet by multiplying the value in the "Qty Fix" column by the value in the "Fixture Watts" column. Annual hours of use values were provided by the client. "Annual kWh" values were computed by the spreadsheet by multiplying values in the "Total Watts" column by values in the "Annual Hrs Use" column. The most important part of this spreadsheet is the information in the "Watts/sf" column, calculated by the spreadsheet by dividing values in the "Total Watts" column by values in the "Net sf" column. Power density (W/sf) values greater than 4 W/sf are printed in bold font and are marked with a "<" symbol to highlight these high values.

Offices with high power densities are prime candidates for lighting retrofits. Note that some of these offices have power densities as high as 6.4 W/sf! The average power density for the group of office areas shown on the spreadsheet is calculated and shown as 4.2 W/sf—about four times what current (1997) lighting design practice is achieving. Totals of square feet, quantity, watts (and kW), and kWh are calculated.

An actual comprehensive audit report is shown on pages 75 through 81.The name of the company and its location have been modified. All the summary tables are shown, but the original data has not been included. Table 3 of the report was compiled using the detailed interior fixture data shown on the spreadsheet in Figure 75. This spreadsheet shows how only six fixture types represent the major portion of the fixtures (86.5%) and how they were identified from the total of 2818 fixtures in the facility. Figure 7.6 shows the Exterior Fixture Schedule spreadsheet and indicates that these fixtures only use 0.1% of the total lighting energy of the facility. For this reason, and because they are operated at off-peak times, upgrades to the exterior lighting were recommended for reducing maintenance costs and to improve safety and security, rather than for energy efficiency.

**TABLE 7.1**

| Fixture Type | Quantity | % | kW | % | kWh | % |
|---|---|---|---|---|---|---|
| Incandescent | 76 | 25% | 16.6 | 28% | 45,402 | 25% |
| Fluorescent | 169 | 57% | 24.3 | 42% | 72,102 | 39% |
| HID | 40 | 14% | 17.0 | 29% | 61,243 | 33% |
| EXIT | 12 | 4% | 0.6 | 1% | 5,256 | 3% |
| Totals | 297 | | 58.5 | | 184,003 | |

RFG Co.
Southern, IN
1995 LIGHTING SURVEY
4-Lp 2x4 Recessed Troffers

| LOCATION | SPACE | NET SF | QTY Fix | FIX TYPE | FIX SIZE | LAMP TYPE | LAMP DESCRIP | LAMP WATTS | LPS/ FIX | BALLAST TYPE | LPS/ BAL | FIXTURE WATTS | TOTAL WATTS | ANNUAL HRS USE | ANNUAL KWH | WATTS/ SF |
|---|---|---|---|---|---|---|---|---|---|---|---|---|---|---|---|---|
| Building 1 | stairwell | 200.0 | 5 | 15 | 50 | 56 | RW | 34 | 4 | 80 | 2 | 164 | 820 | 4,200 | 3,444 | **4.1 <** |
| | Taft | 400.0 | 6 | 15 | 50 | 56 | RW | 34 | 4 | 80 | 2 | 164 | 984 | 4,200 | 4,133 | 2.5 |
| | Blasey | 288.0 | 6 | 15 | 50 | 56 | RW | 34 | 4 | 80 | 2 | 164 | 984 | 4,200 | 4,133 | 3.4 |
| | Sec Off-Lofton | 138.0 | 2 | 15 | 50 | 59 | STD | 40 | 4 | 80 | 2 | 192 | 384 | 4,200 | 1,613 | 2.8 |
| | Exec Conf Rm | 528.0 | 6 | 15 | 50 | 59 | STD | 40 | 4 | 80 | 2 | 192 | 1,152 | 4,200 | 4,838 | 2.2 |
| | Offenbach | 400.0 | 8 | 15 | 50 | 59 | STD | 40 | 4 | 80 | 2 | 192 | 1,536 | 4,200 | 6,451 | 3.8 |
| | Bast | 224.0 | 4 | 15 | 49 | 59 | STD | 20 | 4 | 80 | 2 | 112 | 448 | 4,200 | 1,882 | 2.0 |
| | | | 2 | 15 | 50 | 59 | STD | 40 | 4 | 80 | 2 | 192 | 384 | 4,200 | 1,613 | |
| | Secretary | 310.0 | 6 | 15 | 50 | 59 | STD | 40 | 4 | 80 | 2 | 192 | 1,152 | 4,200 | 4,838 | 3.7 |
| | corridor | 192.0 | 2 | 15 | 50 | 59 | STD | 40 | 4 | 80 | 2 | 192 | 384 | 4,200 | 1,613 | 2.0 |
| | | | 1 | 15 | 50 | 56 | RW | 34 | 4 | 80 | 2 | 164 | 164 | 8,760 | 1,437 | |
| | Mellar | 140.0 | 5 | 15 | 50 | 56 | RW | 34 | 4 | 80 | 2 | 164 | 820 | 4,200 | 3,444 | **5.9 <** |
| | hall to personnel | 160.0 | 4 | 15 | 50 | 56 | RW | 34 | 4 | 80 | 2 | 164 | 656 | 4,200 | 2,755 | **4.1 <** |
| | personnel | 1,736.0 | 46 | 15 | 50 | 56 | RW | 34 | 4 | 80 | 2 | 164 | 7,544 | 4,200 | 31,685 | **4.3 <** |
| | alcove to pers | 185.0 | 4 | 15 | 50 | 56 | RW | 34 | 4 | 80 | 2 | 164 | 656 | 4,200 | 2,755 | 3.5 |
| | storage-personn | 80.0 | 3 | 15 | 50 | 56 | RW | 34 | 4 | 80 | 2 | 164 | 492 | 4,200 | 2,066 | **6.2 <** |
| | Jilton | 144.0 | 4 | 15 | 50 | 56 | RW | 34 | 4 | 80 | 2 | 164 | 656 | 4,200 | 2,755 | **4.6 <** |
| | supply rm | 96.0 | 3 | 15 | 50 | 56 | RW | 34 | 4 | 80 | 2 | 164 | 492 | 4,200 | 2,066 | **5.1 <** |
| | Wilson | 120.0 | 4 | 15 | 50 | 56 | RW | 34 | 4 | 80 | 2 | 164 | 656 | 4,200 | 2,755 | **5.5 <** |
| | S. Smith | 120.0 | 4 | 15 | 50 | 56 | RW | 34 | 4 | 80 | 2 | 164 | 656 | 4,200 | 2,755 | **5.5 <** |
| | Spinder | 80.0 | 2 | 15 | 50 | 56 | RW | 34 | 4 | 80 | 2 | 164 | 328 | 4,200 | 1,378 | **4.1 <** |
| | personnel conf rm | 168.0 | 6 | 15 | 50 | 56 | RW | 34 | 4 | 80 | 2 | 164 | 984 | 4,200 | 4,133 | **5.9 <** |
| | Spronger | 168.0 | 6 | 15 | 50 | 56 | RW | 34 | 4 | 80 | 2 | 164 | 984 | 4,200 | 4,133 | **5.9 <** |
| | next to Spronger | 168.0 | 6 | 15 | 50 | 56 | RW | 34 | 4 | 80 | 2 | 164 | 984 | 4,200 | 4,133 | **5.9 <** |
| | Cestrall | 168.0 | 6 | 15 | 50 | 56 | RW | 34 | 4 | 80 | 2 | 164 | 984 | 4,200 | 4,133 | **5.9 <** |
| | Skulden | 168.0 | 6 | 15 | 50 | 56 | RW | 34 | 4 | 80 | 2 | 164 | 984 | 4,200 | 4,133 | **5.9 <** |
| | physical assess | 90.0 | 1 | 15 | 50 | 56 | RW | 34 | 4 | 80 | 2 | 164 | 164 | 4,200 | 689 | 1.8 |
| | first aid | 80.0 | 2 | 15 | 50 | 56 | RW | 34 | 4 | 80 | 2 | 164 | 328 | 4,200 | 1,378 | **4.1 <** |
| | x-ray | 80.0 | 2 | 15 | 50 | 56 | RW | 34 | 4 | 80 | 2 | 164 | 328 | 4,200 | 1,378 | **4.1 <** |
| | drug screening | 75.0 | 2 | 15 | 50 | 56 | RW | 34 | 4 | 80 | 2 | 164 | 328 | 4,200 | 1,378 | **4.4 <** |
| | first-aid hall | 200.0 | 5 | 15 | 50 | 56 | RW | 34 | 4 | 80 | 2 | 164 | 820 | 4,200 | 3,444 | **4.1 <** |
| | first-aid lobby | 112.0 | 3 | 15 | 50 | 56 | RW | 34 | 4 | 80 | 2 | 164 | 492 | 4,200 | 2,066 | **4.4 <** |
| | first-aid office | 90.0 | 2 | 15 | 50 | 56 | RW | 34 | 4 | 80 | 2 | 164 | 328 | 4,200 | 1,378 | 3.6 |
| | nurse's office | 104.0 | 2 | 15 | 50 | 56 | RW | 34 | 4 | 80 | 2 | 164 | 328 | 4,200 | 1,378 | 3.2 |
| | ER | 192.0 | 4 | 15 | 50 | 56 | RW | 34 | 4 | 80 | 2 | 164 | 656 | 4,200 | 2,755 | 3.4 |
| | nurse's station | 104.0 | 2 | 15 | 50 | 56 | RW | 34 | 4 | 80 | 2 | 164 | 328 | 4,200 | 1,378 | 3.2 |
| | file rm | 144.0 | 4 | 15 | 50 | 56 | RW | 34 | 4 | 80 | 2 | 164 | 656 | 4,200 | 2,755 | **4.6 <** |
| | office & alcove | 128.0 | 5 | 15 | 50 | 56 | RW | 34 | 4 | 80 | 2 | 164 | 820 | 4,200 | 3,444 | **6.4 <** |
| | indust rel's | 132.0 | 4 | 15 | 50 | 56 | RW | 34 | 4 | 80 | 2 | 164 | 656 | 4,200 | 2,755 | **5.0 <** |
| | off adj to 1st aid | 121.0 | 3 | 15 | 50 | 56 | RW | 34 | 4 | 80 | 2 | 164 | 492 | 4,200 | 2,066 | **4.1 <** |
| | corridor | 192.0 | 6 | 15 | 50 | 56 | RW | 34 | 4 | 80 | 2 | 164 | 984 | 4,200 | 4,133 | **5.1 <** |
| | lobby | | 9 | 15 | 50 | 56 | RW | 34 | 4 | 80 | 2 | 164 | 1,476 | 4,200 | 6,199 | |
| | hall | 60.0 | 1 | 15 | 50 | 56 | RW | 34 | 4 | 80 | 2 | 164 | 164 | 4,200 | 689 | 2.7 |
| | meeting rm | 135.0 | 2 | 15 | 50 | 56 | RW | 34 | 4 | 80 | 2 | 164 | 328 | 4,200 | 1,378 | 2.4 |
| | receptionist | 256.0 | 4 | 15 | 50 | 56 | RW | 34 | 4 | 80 | 2 | 164 | 656 | 4,200 | 2,755 | 2.6 |
| TOTALS | SF = | 8,676 | 220 | | | | | | | | | Watts = | 36,600 | kWh= | 154,468 | **4.2** |
| | | | | | | | | | | | | kW= | 36.6 | | | Avg^ |

**FIGURE 7.4** Sample spreadsheet for a comprehensive Green Lights audit. Offices with high power densities, prime candidates for lighting retrofits, are marked with a "<" symbol

An Effective Lighting Solutions, Inc. Survey & Analysis for X Y Z Corporation

Anytown, U.S.A.
Manufacturing Facility

LIGHTING SURVEY

INTERIOR FIXTURE SCHEDULE

| Interior Fixtures | Fixture Code | Fixture Description | Lamp Description | Lamp Watts | Lamps/ Fixture | Ballast Type | Lamps/ Ballast | Fixture Watts | Fixture Quantity | % of Total |
|---|---|---|---|---|---|---|---|---|---|---|
| Incandescent | II-1 | Lamp Holder | PAR-38 | 150 | 1 | | | 150 | 107 | 3.8% |
| | II-2 | Recessed Shower Light | A-lamp | 100 | 1 | | | 100 | 8 | 0.3% |
| | II-3 | Restroom Round Light | A-lamp | 100 | 1 | | | 100 | 2 | 0.1% |
| | II-4 | Quartz Flood | Double ended | 300 | 1 | | | 300 | 12 | 0.4% |
| | II-5 | Downlight | 75R30 | 75 | 1 | | | 75 | 7 | 0.2% |
| Fluorescent | IF-1 | 2x4 Lensed Troffer | F40/LW/EW | 34 | 4 | Std Mag | 2 | 156 | 789 | 28.0% |
| | IF-2 | 16' Industrial HO | F96T12HO/EW | 95 | 2 | Std Mag | 2 | 219 | 858 | 30.4% |
| | IF-3 | 8' Industrial HO | F96T12HO/EW | 95 | 2 | Std Mag | 2 | 219 | 349 | 12.4% |
| | IF-4 | 8' Industrial HO | F96T12HO/EW | 95 | 1 | Std Mag | 1 | 120 | 198 | 7.0% |
| | IF-5 | 4' Industrial HO | F48T12/CW/HO | 60 | 1 | Std Mag | 1 | 80 | 7 | 0.2% |
| | IF-6 | 8' Industrial T-8 | F32T8 | 32 | 2 | 2-lp ELB | 2 | 61 | 3 | 0.1% |
| | IF-7 | 2x4 Troffer | F32T8 | 32 | 4 | 4-lp ELB | 4 | 121 | 42 | 1.5% |
| | IF-8 | 2x2 Troffer | F40/6U | 40 | 2 | Std Mag | 2 | 96 | 74 | 2.6% |
| | IF-9 | 1x4 Fixture w/Gold louver | F40T12/EW | 34 | 2 | Std Mag | 2 | 78 | 15 | 0.5% |
| | IF-10 | 4'-1 lp T12 | F40T12/EW | 34 | 1 | Std Mag | 1 | 48 | 3 | 0.1% |
| | IF-11 | 4'-2 lp T12 | F40T12/EW | 34 | 2 | Std Mag | 2 | 78 | 137 | 4.9% |
| | IF-12 | 4'-2 lp HO | F48T12/CW/HO | 60 | 2 | Std Mag | 2 | 128 | 10 | 0.4% |
| | IF-13 | 4'-4 lp T12, wrap-around | F40/LW/EW | 34 | 4 | Std Mag | 2 | 156 | 2 | 0.1% |
| | IF-14 | 2'-2 lp T12 | F20T12/CW | 20 | 2 | Std Mag | 2 | 58 | 11 | 0.4% |
| High Intensity Discharge | IH-1 | CYL w/Fluted Sides | MH | 175 | 1 | Std Mag | 1 | 220 | 34 | 1.2% |
| | IH-2 | GE Versabeam ("teardrop" | MH | 250 | 1 | Std Mag | 1 | 296 | 8 | 0.3% |
| | IH-3 | GE Lowmount | HPS | 250 | 1 | Std Mag | 1 | 308 | 38 | 1.3% |
| | IH-4 | Coneshape | HPS | 150 | 1 | Std Mag | 1 | 195 | 30 | 1.1% |
| | IH-5 | Open, Bell Shape,Metal H | MH | 400 | 1 | Std Mag | 1 | 460 | 50 | 1.8% |
| | IH-6 | Holophane Prismalume | HPS | 400 | 1 | Std Mag | 1 | 475 | 8 | 0.3% |
| | IH-7 | 1X1 HPS | HPS | 50 | 1 | Std Mag | 1 | 83 | 10 | 0.4% |
| | | | | | | | | | 2,818 | |

**FIGURE 7.5** Spreadsheet showing interior lighting fixture data, used for analysis for the comprehensive lighting audit report reproduced in this chapter.

An Effective Lighting Solutions, Inc. Survey & Analysis for X Y Z Corporation

Anytime, U.S.A.
Manufacturing Facility

1996 LIGHTING SURVEY

EXTERIOR FIXTURE SCHEDULE

| Exterior Fixtures | Fixture Code | Fixture Description | Lamp Description | Lamp Watts | Lamps/ Fixture | Ballast Type | Lamps/ Ballast | Fixture Watts | Fixture Qty | Total Watts | Off-peak Hours of Use | Total kWh/yr |
|---|---|---|---|---|---|---|---|---|---|---|---|---|
| Incandescent | EI-1 | Lamp Holder | PAR-38 | 150 | 2 | | 2 | 300 | 3 | 900 | 4,380 | 3,942 |
| | EI-2 | Hazardous rated | A-lamp | 100 | 1 | | 1 | 100 | 1 | 100 | 4,380 | 438 |
| | EI-3 | Downlight (DL) | PAR-38 | 150 | 1 | | 1 | 150 | 2 | 300 | 4,380 | 1,314 |
| HID | EH-1 | Cobrahead | HPS | 150 | 1 | Std Mag | 1 | 185 | 5 | 925 | 4,380 | 4,052 |
| | EH-2 | Cobrahead | Mercury | 175 | 1 | Std Mag | 1 | 205 | 7 | 1,435 | 4,380 | 6,285 |
| | EH-3 | Holophane Petrolux | HPS | 100 | 1 | Std Mag | 1 | 128 | 10 | 1,280 | 4,380 | 5,606 |
| | EH-4 | Large floodlight | Mercury | 175 | 1 | Std Mag | 1 | 205 | 11 | 2,255 | 4,380 | 9,877 |
| | EH-5 | Small flood/wallpac | HPS | 100 | 1 | Std Mag | 1 | 128 | 16 | 2,048 | 4,380 | 8,970 |
| | EH-6 | Small wallpack | Mercury | 175 | 1 | Std Mag | 1 | 205 | 15 | 3,075 | 4,380 | 13,469 |
| | EH-7 | Yard light | Mercury | 175 | 1 | Std Mag | 1 | 205 | 1 | 205 | 4,380 | 898 |
| | EH-8 | Flag light | HPS | 50 | 1 | Std Mag | 1 | 83 | 1 | 83 | 4,380 | 364 |
| | EH-9 | Large canopy DL | Mercury | 175 | 1 | Std Mag | 1 | 205 | 10 | 2,050 | 4,380 | 8,979 |
| | | | | | | | | | 82 | | | |
| | | | | | | | | | kW = | 14.7 | kWh = | 64,193 |
| | | | | | | | | | | | per-cent of tota= | 0.1% |

**FIGURE 7.6** Spreadsheet showing exterior lighting fixture data, used for analysis for the comprehensive lighting audit report reproduced in this chapter.

**XYZ CORPORATION**

**Anytown, U.S.A.**

**Facility**

**Lighting Survey**

**Report**

**May 31, 1996**

**Prepared by**

**Effective Lighting Solutions, Inc.**

**1015 E. Main St.**

**Columbus, OH 43205**

# 1.0 EXECUTIVE SUMMARY

## 1.0 Study Objective

Effective Lighting Solutions, Inc., was contracted by XYZ Corporation to conduct a comprehensive lighting survey of the Anytown plant. The purpose of the survey was to inventory all lighting systems and to recommend cost-effective lighting upgrades. All retrofit opportunities were analyzed for economic feasibility using the simple payback method.

## 1.2 Scope of Study and Report

The study was conducted in two phases. Phase one involved a thorough inspection of the Anytown facility. On March 23–24, 1995, all lighting systems of the Anytown facility were surveyed by a team of trained auditors. Building sketches were reviewed and building data was collected. Details of building occupancy patterns, operational procedures, and lighting maintenance were provided by XYZ personnel. Phase two involved analyzing power bills, preparing the inventory of lighting equipment, identifying and evaluating upgrade opportunities, and performing economic analyses.

There are two main sections in this report: Introduction and Lighting Systems. The introduction provides an overview of the facility and documents current lighting system energy consumption patterns. The lighting systems section presents the field survey findings, examines retrofit alternatives, and makes recommendations for action by XYZ.

### 1.3 Results

Of the six fixture types identified for potential upgrading, only two were found to be good candidates for cost-effective energy-saving retrofits.

A simple bulb change from a standard incandescent PAR flood to a tungsten-halogen PAR will save $2250 each year and pay back the investment in 4 months. This retrofit is identified as ECM-1 (ECM stands for "energy conservation measure").

A retrofit of the 2-lamp HO fixtures will save $15,780 annually and will pay back the investment in 4.3 years. This retrofit is identified as ECM-3.

## 2.0 INTRODUCTION

### 2.1 Facility Description

A facility of XYZ Corporation, this manufacturing facility is located at 1200 Pennsylvania Avenue in Anytown, U.S.A. The facility is dedicated to the manufacture of quality products for the building industry. The facility totals more than 376,000 square feet.

### 2.2 Historical Demand and Electrical Energy Consumption Data

Historical data on facility electrical demand, energy and costs were supplied by XYZ facilities personnel. A summary of this data for the time period February 1994 to January 1995 is presented in Table 1. (*Note: At the request of the client, Table 1 and the graph and pie chart mentioned in this section have been excluded from this sample report.*) The table shows the demand (kW) and energy (kWh) for the total facility for a 12-month period. A graph of monthly usage and costs follows Table 1. Electrical cost data is presented in the pie chart following the graph of monthly usage and costs. Billing data was included for the purpose of calculating savings in the same way that electric bills are paid for the facility. Currently, the Anytown facility purchases electrical power to operate its processes to produce building materials. In addition, electrical power is used for building lighting. The lighting systems account for 3.4% of the electrical energy and 7.1% of the minimum power demand purchased for the facility. Table 2 below summarizes the power and electrical energy used for the Anytown lighting systems.

**TABLE 2**

| Interior | Exterior |
|---|---|
| On-peak lighting demand:<br>515 kW | Off-peak lighting demand:<br>14.7 kW |
| % of minimum peak demand:<br>7.1% | |
| Annual lighting energy:<br>4,032,380 kWh | Annual lighting energy:<br>64,193 kWh |
| % Total electrical energy (kWh):<br>3.4 % | % Total electrical energy (kWh):<br>0.1 % |
| Incandescent, fluorescent, HID<br>(Table 7.3) | HID—mercury and high-pressure<br>sodium |

## 3.0 LIGHTING SYSTEMS

### 3.1 Introduction

Lighting functions to provide safety, security, and visual task recognition. Light contributes to the psychological and physiological well-being of the building occupants. Electric lighting, while providing these functions, consumes electrical energy, and lighting systems require maintenance. The quantity of light in a space likewise has an effect on the occupants. Too much light wastes energy and contributes to unnecessary maintenance, while too little light negatively impacts the occupants and affects employee productivity. The goal of any lighting system is to maintain a balance between function and cost so that the occupants and the owner/operators may both benefit. The recommendations that follow will use this guideline and, where appropriate, will offer suggestions for overlit areas, alternative light sources, controls, and maintenance.

The discussions that follow will provide a current status of the lighting systems and will detail suggested changes targeted at reducing electrical demand and energy (and its attendant power plant emissions) and at improving lighting system performance.

## 3.2 Lighting Systems

A summary of all the lighting systems for the XYZ facility is provided in Table 3.

**TABLE 3**

| Lighting System | Fixture Quantity | % of Total |
|---|---|---|
| Incandescent | 136 | 5 % |
| Fluorescent | 2504 | 89 % |
| HID | 178 | 6 % |
| Total | 2818 | 100 % |

**1. Exterior Systems**

The exterior lighting systems account for only 0.1% of the total electrical energy used at the Anytown plant. For this and for reasons of limited hours of use and a favorable electric rate, no energy conservation measures for the exterior systems were found to be cost-effective.

Mercury-vapor fixtures are still in use as part of the exterior lighting system. New, high-pressure sodium fixtures are recommended. Although there is no energy incentive for these conversions, they should be done to reduce maintenance costs. High-pressure sodium fixtures should be chosen with a lower wattage rating than the mercury fixtures they replace.

Table 4 (*Figure 7.6*) shows a detailed inventory of the exterior lighting fixtures and the estimated electrical energy consumption.

**2. Interior Systems**

A summary of the major interior lighting systems in the Anytown plant is shown in Table 5. Six (6) fixture types represent 86.5% of the total interior lighting fixtures in the facility, accounting for 6% of the minimum peak demand and 83.7% of the lighting energy (kWh/yr).

## 3.3 State of Office Lighting Systems

1. Lighting in most of the office spaces is provided by fixture code IF-1, the 2 × 4 lensed, 4-lamp recessed troffer, equipped with four 34-watt reduced-wattage T12 lamps and standard magnetic ballasts.
2. Many fixtures are in poor condition, with discolored polystyrene lenses, broken lenses, and lenses falling out of the fixtures. These fixtures should either be repaired or replaced with new fixtures using state-of-the-art T8 lamps and electronic ballasts.
3. Some conversions to the more efficient T8 lamps and electronic ballasts have been started.

## 3.4 State of Industrial Lighting Systems

1. The use of HID and HO (high-output) fluorescent fixtures is an effective combination of lighting systems for lighting industrial tasks that should be continued. Although there is a wide variation of HID fixture types in use, there is little opportunity for improvement with the HID systems to make them more efficient. Where good color rendition is **not** a requirement, the use of high-pressure sodium lighting is the preferred choice, due to its high efficiency.
2. Lamps are changed on a spot basis when they burn out. A higher overall light level could be achieved from the HO fluorescent lighting if scheduled replacement of these high lumen depreciation lamps was implemented in large groups, based on the desired light level point on the depreciation curve. The lamp lumen depreciation curve for HO lamps is rather steep and indicates that less than 70% of the initial light level is available at the end of lamp life.
3. The use of higher CRI (color rendering index) standard-wattage lamps should be explored for inspection areas along with the use of electronic ballasts that would produce less flicker and result in fewer worker complaints. An alternative would be to use multiple T8 lamps and electronic ballasts.
4. There is a need for improved lighting in the C.R. area of the facility. The critical visual task here is quality inspection. The bare HO lamps cause glare, and the color rendition is poor, resulting in a lack of contrast and poor visibility. A lensed fixture providing good optical control, such as the Holophane Prismawrap, employing high CRI lamps operated at high frequency by an electronic ballast, will provide the

**TABLE 5**

| Fixture Code | Fixture Description | Lamp Description | Fixture Quantity | % of Total | kW | % of Minimum Demand | kWh | % of Lighting kWh |
|---|---|---|---|---|---|---|---|---|
| II-1 | Single lamp holder | PAR-38 flood or spot | 107 | 3.8% | 16.1 | 0.2% | 140,600 | 3.5% |
| IF-1 | 2x4 lensed 4-lamp troffer | 4' 34-W reduced-wattage fluorescent | 789 | 28.0% | 123.1 | 1.7% | 665,700 | 28.9% |
| IF-2 | 16' 2-lamp industrial | 8' 95-W reduced-wattage HO fluorescent | 858 | 30.4% | 187.9 | 2.6% | 1,627,100 | 40.4% |
| IF-3 | 8' 2-lamp industrial | 8' 95-W reduced-wattage HO fluorescent | 349 | 12.4% | 76.4 | 1.1% | 665,400 | 16.5% |
| IF-4 | 8' 1-lamp industrial | 8' 95-W reduced-wattage HO fluorescent | 198 | 7.0% | 21.1 | 0.3% | 185,000 | 4.6% |
| IF-11 | 4' 2-lamp various styles | 4' 34-W reduced-wattage fluorescent | 137 | 4.9% | 10.7 | 0.1% | 92,700 | 2.3% |
| | | Totals | 2438 | 86.5% | 435.3 | 6.0% | 3,376,500 | 83.7% |

improvement needed. Implementing the details of this upgrade requires professional design assistance. Effective Lighting Solutions has the capabilities to provide this assistance.

## 3.5 Field Observations

1. A detailed lighting survey was conducted to identify the fixtures and lamps in use. The lighting systems are primarily composed of fluorescent and metal-halide fixtures, with a few high-pressure sodium fixtures. A breakdown, showing each type of fixture used in each area, electrical wattages, demand, and annual energy consumption, is shown in the tables of the section labeled "Detailed Survey Results." (*Note: This section of the actual report has not been included in this sample report.*)
2. Facility occupancy is regular, and factory operating hours are based on full-time production schedules. Hours of use for labs and high-occupancy production offices is estimated at 5000 hours per year; administration offices are calculated to be occupied 3000 hours per year. The exterior areas are on for an average of 12 hours per night, calculated at 4380 hours per year.
3. A newer lighting practice has demonstrated effectiveness and should be continued and their use expanded where possible:
   A. The use of reduced wattage HO lamps are appropriate and should be continued.
4. Some lighting practices should be discontinued in favor of more effective practices:
   A. Some of the 8-foot HO fixtures are mounted too high to be effective. At mounting heights above 10 feet, HID fixtures would be more effective.
   B. In some cases, it appears that the choice of HID fixtures is based on first cost only. Consideration should be given to low-loss ballasts, which can pay for themselves over their lifetime.
5. The following new lighting practice should be initiated:
   A. Some general purpose incandescent lamps are in use throughout the facility. Cost-effective tungsten-halogen replacements for standard-wattage A lamps are recommended and are shown in Table 6.

**TABLE 6**

| Old | New |
|---|---|
| 60 watt (60 A) | 42MB/CAP |
| 75 watt (75 A) | 52MB/CAP |
| 100 watt (100 A) | 72MB/CAP |

These lamps will save \$6–\$9 per lamp in energy costs and have 3500-hour rated lamp life, compared with 750- to 1000-hour life of the standard A lamps now being used. You can use these tungsten-halogen lamps anywhere standard incandescent lamps are now being used. They are also available in 130-volt versions for difficult-to-reach areas.

## 3.6 Summary of Lighting Energy Conservation Measures (ECMs)

Table 7 presents a summary of Lighting ECMs.

## 3.7 Lighting ECM$ Details (Table 7)

### ECM-1

The simplest retrofit to implement and the least expensive is a bulb change. Changing from the standard 150-watt PAR-38 incandescent to a 75-watt PAR-38 tungsten-halogen lamp will provide the same beam lumens (same quantity of light) for 50% of the power and energy. Since many of these lamps are used in process areas that operate on a constant burn schedule, the payback is 4 months on energy costs alone. The cost to implement this ECM is calculated on the basis of lamp cost only, the labor being provided by in-house electrical maintenance staff. Because of the longer rated lamp life of the retrofit lamp, the time between lamp changes will decrease, decreasing maintenance costs.

### ECM-2

An upgrade of the 4-lamp T12 fluorescent fixtures to 4-lamp T8 lamps driven with a 4-lamp electronic ballast was explored. Short hours of use and a favorable electric rate cause a simple payback of 7.1 years. The estimated cost to implement this ECM is \$42,600 for retrofitting 789 fixtures. The energy cost savings per year are calculated to be \$6000. This measure is **not** cost-effective at this time.

**TABLE 7**

| Fixture Code | Fixture Description | Lamp Description | ECM | ECM Description | kW Savings per Month | kW Savings per Year | Energy Cost Savings per Year | SPB (Years) |
|---|---|---|---|---|---|---|---|---|
| II-1 | Single lamp holder | PAR-38 flood or spot | 1 | Change bulb to 75PAR38/ CAP | 8.0 | 70,300 | $2,250 | 0.3 |
| IF-1 | 2x4 lensed 4-lamp | 4' 34-W reduced-wattage fluorescent | 2 | Retrofit fixture w/ 4 T8 lamps + electronic ballast | | | NOT cost-effective | 7.1 |
| IF-2 | 16' 2-lamp industrial | 8' 95-W reduced-wattage HO fluorescent | 3 | Retrofit fixture w/ HO electronic ballast | 40.3 | 349,200 | $11,200 | 4.3 |
| IF-3 | 8' 2-lamp industrial | 8' 95-W reduced-wattage HO fluorescent | 3 | Retrofit fixture w/ HO electronic ballast | 16.4 | 142,700 | $4,580 | 4.3 |
| IF-4 | 8' 1-lamp industrial | 8' 95-W reduced-wattage HO fluorescent | 4 | NO retrofit available | | | | |
| IF-11 | 4' 2-lamp, various styles | 4' 34-W reduced-wattage fluorescent | 5 | Retrofit fixtures w/ 2 T8 lamps + electronic ballast | | | | |
| | | Totals | | | 64.8 | 562,200 | $18,030 | 3.8 |

**ECM-3**

Several options for upgrading the 8- and 16-foot HO 2-lamp fixtures were investigated. These included conversion to slimline lamps and reflectors and retrofit with reflectors and T8 lamps. The only option that is cost-effective is the conversion to an HO electronic ballast. The estimated cost to change to the HO electronic ballast is $48,000 for the 858 16-foot 2-lamp fixtures and $19,600 for the 349 8-foot 2-lamp fixtures. Annual energy cost savings for the 16-foot fixtures calculates to $11,200 and are $4580 for the 8-foot fixtures.

If this option is implemented, the following specifications should be followed when selecting the HO electronic ballast: Must operate 95-watt HO lamps with a maximum ANSI input wattage of 172 watts, minimum ballast factor of 0.84, maximum current crest factor of 1.6, and maximum THD of 20%.

Should XYZ decide not to implement this ECM at this time, there is a recent development that may prove to be cost-effective in the future. That development is a T8 HO retrofit lamp to replace the HO lamp. A change to a T8 HO electronic ballast is required. So far, only one lamp company has announced the availability of this lamp, and no ballast is yet available in production quantities. Should other lamp and ballast manufacturers come out with similar products, making this option a viable alternative (this company does not consider single-source products), it should be reinvestigated for future use in the Anytown facility.

**ECM-4**

A literature search for a single-lamp HO electronic ballast to use in fixture code IF-4 shows none to be available. With only 198 fixtures of this type in the plant, it is **not** considered a cost-effective retrofit. Theoretically, tandem ballasting could be implemented where there is an adjacent fixture in close proximity, but practical wiring considerations would make the installation cost too high to provide the required payback.

**ECM-5**

A retrofit of the 2-lamp, 4-foot T12 fluorescent fixtures to 2-lamp T8 lamps driven with a 4-lamp electronic ballast was analyzed. The simple payback calculates to 9.1 years, due to short hours of use and a favorable electric rate. The estimated cost to implement this ECM is $6850 for retrofitting 137 fixtures. The annual energy cost savings per year calculates to $750. This measure is **not** cost-effective at this time.

## 3.8 Economic Evaluation

The payback calculations were performed using the actual electric bill rate structure that XYZ is billed, using 1994 bills to calculate the savings based on peak demand savings and on-peak and off-peak kWh savings.

## 3.9 Summary

The comprehensive lighting survey conducted for the XYZ Corporation Anytown Plant shows that there are two cost-effective lighting retrofits at this time: ECM-1 and ECM-3. The total cost to implement these measures requires an investment of $68,300, with an annual energy savings of $18,030, resulting in a simple investors return on investment (ROI) of 26.3%.

Several recommendations to improve lighting practices have been presented and an upgrade of the lighting for a critical inspection task in the C.R. area has been described.

Chapter 8

# Lighting Fact Sheets

## Fact Sheet 1: Lighting Basics

### Quantity of Light

The total light from a light source is measured in lumens. Lamps are labeled with measured lumen ratings and efficiency ratings (efficacy). In the English system, the quantity of light that falls on a work surface is measured in footcandles. One footcandle (FC) is equal to one lumen per square foot of area. In the SI system, the lux is the measure of illuminance. One lux is equal to one lumen per square meter of area. One FC is equal to about 10 lux.

The catalog rating for lamps is measured in initial illuminance, which defines the light output after 100 burn hours. Lamps lose their ability to produce light over time. This characteristic is called lamp lumen depreciation (LLD). The light output at 40% life is measured in mean (or design) lumens. Lamp lumen depreciation is defined as the ratio of mean to initial lumens.

An illuminance meter is a useful tool for measuring the amount of light in work spaces. It is important to understand that the footcandle measure indicates only a level of illuminance. It does not measure the amount of energy needed to produce that level of light and does not measure the quality of the light produced.

How much light does it take for people to see? We know from experience that it takes more light to see small objects and perform tasks of low contrast than it does to see and perform larger, higher-contrast tasks. As our eyes age, they require more light. In fact, a 50-year-old gets about half as much light on the back of the eye, at the retina, as does a 20-year-old.

How much light is required to illuminate work spaces or to perform tasks? The Illuminating Engineering Society of North America (IESNA) provides light level (illuminance) recommendations. These recommendations are consensus standards that have been agreed upon by lighting professionals.

The IESNA *Lighting Handbook*, 8th edition, contains the latest illuminance recommendations for different types of spaces and for various tasks. Use of these recommendations will avoid overlighting. Overlighting spaces not only wastes energy but also reduces lighting quality.

### Quality of Light

Lighting quality is complex and there is no single unit of measure for it that is comparable to illuminance (FC), the unit used to designate lighting quantity. Lighting quality considers visual aspects that are highly subjective and not easily quantified. We can, however, understand certain quality characteristics.

**Veiling reflections** detract from lighting quality by obscuring task details by reducing contrast. Also called reflected glare, veiling reflections are most noticeable from luminaires located in front and above the viewing task.

**Glare** is defined as the sensation produced by "brightness" within the visual field that is sufficiently greater than the "brightness" to which the eye is adapted such that a loss of visual performance occurs. In building interiors, discomfort glare is annoying and diminishes lighting quality by interfering with employee productivity more than with actual visual performance.

Glare can be controlled by controlling the light level and by blocking the direct view of bare lamps. Glare-reducing features of luminaires control light output. These include louvers, such as are found in deep-cell parabolic luminaires, or low-glare acrylic lenses, that reduce surface brightness at high viewing angles. The use of indirect lighting is perhaps the best way to reduce glare in office interiors and is recommended in offices with intensive computer use.

**Lamp color** also affects lighting quality (see Fact Sheet 3). Recommendations regarding pleasant combinations of lamp color temperature and illuminance are changing and are best left to building occupant preferences. However, we do know that when lamps of high color rendering are used, illuminance may be lowered without sacrificing brightness or visual clarity. When upgrading from cool-white lamps to higher CRI T8 lamps, this effect will be noticed by employees, who may complain that the new lamps are "too bright."

**Lamp flicker** can also reduce lighting quality. The predominant source of lamp flicker is from fluorescent lamps operated on magnetic ballasts. The lamps turn on and off 120 times per second, producing distraction, eyestrain, and headaches. Flicker is especially noticeable at high light levels, such as those used in industrial inspection lighting. Electronic ballasts that operate fluorescent lamps at high frequency can reduce flicker to an imperceptible level.

To improve lighting quality, it is important to balance office lighting for visual performance and for visual comfort. Too much contrast will cause workers to be restless. Too little contrast (flat lighting) causes a loss of detail and makes things appear dull. In retail lighting applications, flat lighting lacks the power to attract, and thus sales may be lower under such conditions. Balanced lighting is achieved by avoiding both overdiffuse and too strongly directional lighting.

**Visual comfort probability** (VCP) is a rating of lighting systems that is expressed as a percentage of people who, when viewing from a specified location and in a specified direction, will find the lighting system acceptable in terms of discomfort glare. The IESNA minimum recommendation for office interiors is 70, and the recommendation for computer applications is 80.

## Fact Sheet 2: Lamp and Bulb Identification

**Lamp** is the term lighting professionals use for what are commonly referred to as bulbs and tubes. Identifying different lamp types is a matter of learning the physical shapes, sizes, and bases of bulbs and tubes of the different lamp families. All lamps are identified by a designation that discloses information concerning both physical dimensions and operating characteristics. The designation of lamps is one or two letters, followed by a number. The letters indicate the shape and the number indicates the diameter of the lamp, measured in eighths of an inch. One way to become familiar with lamps and bulbs is to review lamp manufacturer's catalogs.

### Incandescent Bulbs

Figure 8.1 shows a wide variety of incandescent bulbs. Tungsten-halogen (T-H) and quartz lamps are shown in Figure 8.2.

### Fluorescent Tubes

Lamps designated F are fluorescent. The T descriptor designates a tubular shape. A common lamp designation is F40T12/CW/RS. Here is the interpretation:

The number 40 is the lamp wattage. (Note that reduced-wattage lamps are designated SS or WM or EW, depending on the manufacturer. Reduced-wattage lamps are rated 34 watts, although they have a 40 in the catalog number.) The T12 means that this lamp is tubular in shape and is 12 eighths of an inch in diameter. CW designates the phosphor color (cool white), and RS identifies the tube as a rapid-start lamp.

Figure 8.3 shows common linear fluorescent tubes, ranging from T2 axial (a) to circleline lamps (n and o).

### Compact Fluorescent (CFL) Bulbs

Figure 8.4 shows a few of the common CFLs.

### High-Intensity Discharge (HID) Bulbs

Figure 8.5 shows three common HID bulbs. The three bulbs shown have mogul bases, although some metal-halide and high-pressure sodium (HPS) lamps are available in medium bases.

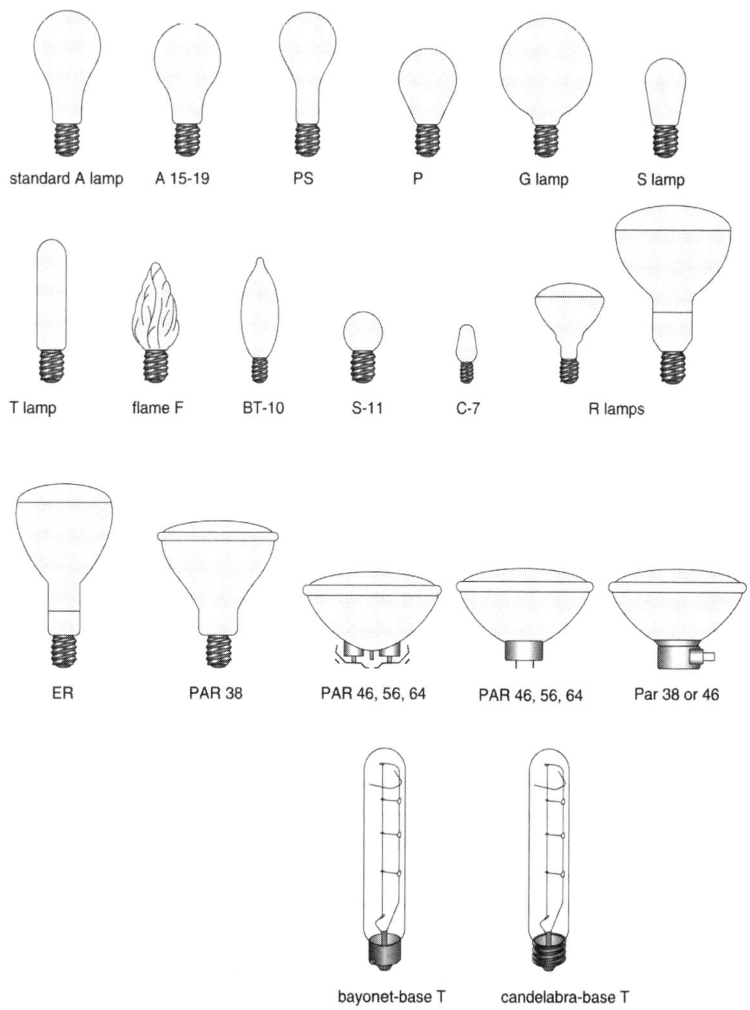


**FIGURE 8.1** A wide variety of incandescent bulbs. A few notes about some of the bulbs pictured: The standard A lamp is the incandescent "arbitrary" shape bulb commonly used in residential and commercial lighting. PS = pear shape, G = globe, T = tubular, R = reflector, ER = ellipsoidal reflector, and PAR = parabolic aluminized reflector. The boyonet-base T and the candelabra-base T are used in EXIT lights, the S-11 is used for older signs and for night-lights, and the C-7 is used as a night-light and as a Christmas tree bulb.

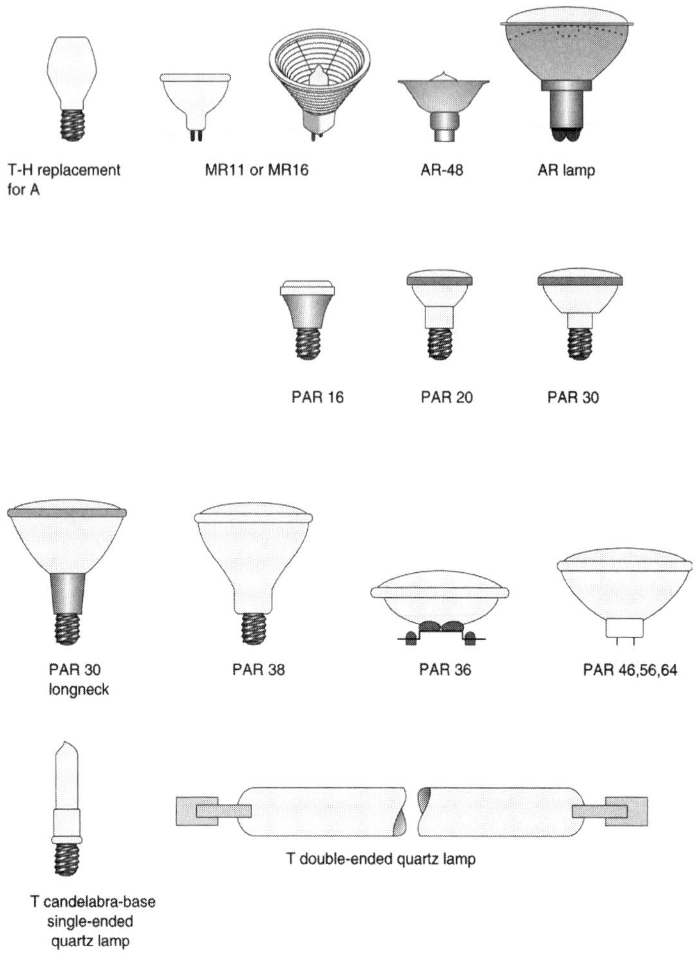


**FIGURE 8.2** A variety of tungsten halogen (T-H) and quartz lamps. Note that tungsten-halogen replacements for A lamps are sometimes called mid-break bulbs. MR stands for "mirrored reflector."

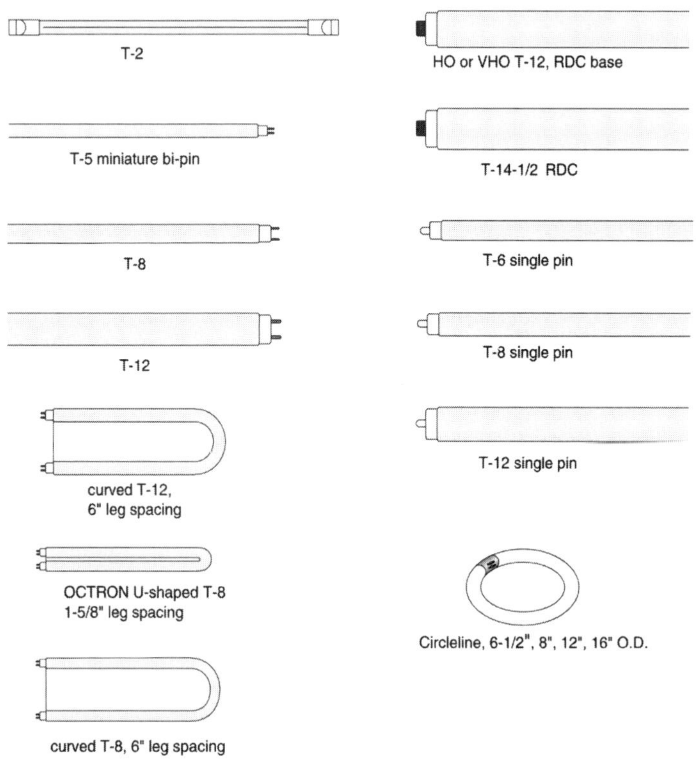


**FIGURE 8.3** Common linear fluorescent tubes. Note that RDC stands for "recessed, double-contact."

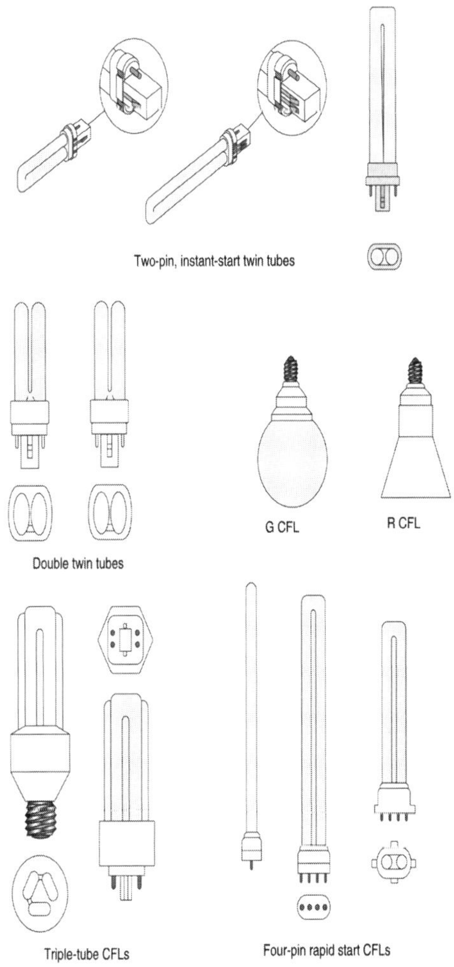


**FIGURE 8.4** Some common compact fluorescent (CFL) bulbs. The two-pin, instant-start twin tubes and the double twin tubes use external ballasts. The G, R, and triple-tube CFLs shown are single-piece CFLs with built-in ballasts (sometimes such lamps are called integrally ballasted or self-ballasted). These one-piece CFLs are used to replace incandescent lamps. The four-pin rapid-start lamps also require an external ballast.

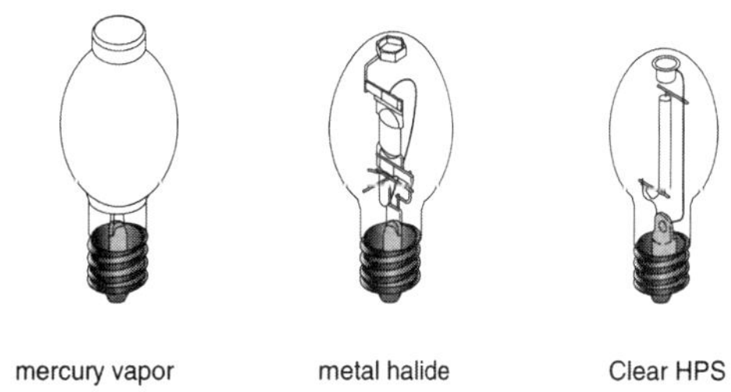


**FIGURE 8.5** Three common high-intensity discharge (HID) bulbs.

## Fact Sheet 3: Lamp Color

In order to understand the role lamp color plays in lighting it is important to understand the relationship between light and color. Light is the radiant energy produced by a light source. Light may come to the observer's eye directly, or it may be reflected by an object. Color is the interaction of the light source, the reflection of light from an object, and the human visual system's ability to discern the color of light. Color cannot be perceived without light. Different light sources emit different wavelengths of light, affecting the appearance of colored objects. Two measures have been developed and are commonly used to describe the color characteristics of lamps. These measures are color temperature and color rendering index.

### Color Temperature

The color temperature of a lamp is described in terms of its appearance to the eye—whether it appears "warm" or "cool." It is measured on a Kelvin scale, ranging from 1500K, which appears red-orange, to 9000K, which appears blue. Light sources lie somewhere between these two, with those of higher color temperature (4100K) appearing "cool" and those of lower color temperature (3000K) appearing "warm." Note that these associations may be confusing (the higher the color temperature, the "cooler" the appearance to the eye, and vice versa). For a more detailed discussion of color temperature, refer to texts that describe how the color appearance measure is related to a reference blackbody radiator. Standard colors for various lamps are shown in Table 8.1.

### Color Rendering Index (CRI)

Color temperature describes the actual appearance of the light being produced by a lamp, but it provides no indication of how the lamp will affect the color appearance of objects. Color rendition describes the effect a light source has on the appearance of

**TABLE 8.1** Standard Colors and CRI Values of Lamps

| Lamp Description | Color Temperature (CCT in K) | Color Rendering Index (CRI) |
|---|---|---|
| 250 W HPS | 2100 | 21 |
| 250 W HPS—deluxe | 2200 | 65 |
| TT or DTT Compact Fluorescent | 2700 | 81 |
| 27 U Fluorescent | 2700 | 82 |
| Incandescent Fluor (IF) | 2750 | 89 |
| 100 W Incandescent | 2900 | 100 |
| Deluxe Warm White (WWX) Fluorescent | 2950–3000 | 74–79 |
| Warm White (WW) Fluorescent | 3000 | 52–53 |
| RE30/70 or Spec 30 Fluorescent | 3000 | 70–75 |
| Soft White Fluorescent | 3000 | 79 |
| RE30/80 or AX30 Fluorescent | 3000 | 80 |
| 30 U Fluorescent | 3000 | 85 |
| Coated 3K Metal Halide | 3200 | 70 |
| Low-Wattage Metal Halide | 3200 | 75 |
| Coated Mercury | 3300 | 52 |
| White (W) Fluorescent | 3450–3500 | 57–58 |
| RE35/70 or Spec 35 Fluorescent | 3500 | 70–73 |
| RE35/80 or AX35 Fluorescent | 3500 | 80–85 |
| 35 U Fluorescent | 3500 | 85 |
| Natural (N) Fluorescent | 3400–3600 | 81–86 |
| Coated Metal Halide | 3600–4000 | 70 |
| Super Metal Halide (Coated) | 3800–4000 | 70 |
| Deluxe Mercury | 3900–4000 | 45–50 |
| Clear Metal Halide | 4000–4400 | 65 |
| RE41/70 or Spec 41 Fluorescent | 4100 | 70 |
| RE41/80 or AX41 Fluorescent | 4100 | 80 |
| 41 U Fluorescent | 4100 | 85 |
| Deluxe Cool White (CWX) Fluorescent | 4100–4200 | 89 |
| Lite White (LW) Fluorescent | 4100–4150 | 48–51 |
| Cool White (CW) Fluorescent | 4100–4200 | 62 |
| RE50/80 or AX50 Fluorescent | 5000 | 80–85 |
| 50 U Fluorescent | 500 | 85 |
| C50 or Colortone 50 Fluorescent | 5000 | 90–92 |
| Clear Mercury | 5710–5900 | 15–22 |
| Daylight Fluorescent | 6300–6500 | 76–79 |
| Cool Green Fluorescent | 6450 | 70 |
| C75 or Colortone 75 Fluorescent | 7500 | 92–95 |

colored objects. The color rendering capability of a lamp is measured as the CRI. The higher the CRI, the less distortion of the object's color by the lamp's light output. The scale used ranges from 0 to 100. A CRI of 100 indicates that there is no color shift as compared to a reference source; thus, the lower the CRI, the more pronounced the shift may be. Generally, the higher the CRI lamp rating, the more natural the appearance of room decoration and furnishings (there is low color distortion). It is important to understand that the reference source and therefore the CRI scale is different at different color temperatures. For this reason, CRI values should only be compared between lamps of similar color temperature. CRI values for various lamps are shown in Table 8.1.

# Fact Sheet 4: Ballasts

Ballasts are required to start and operate discharge lamps, both fluorescent and HID. All fluorescent ballasts perform two functions: they provide the right voltage to start the arc discharge, and they regulate the lamp current to stabilize light output. In rapid-start ballasts, a third function is to provide the energy to heat the electrodes (cathodes).

## Types of Ballasts

The ballasting function has traditionally been reliably accomplished with electromagnetic units that consist of a core of magnetic steel laminations, surrounded by two copper or aluminum coils. Concerns for lighting efficiency have prompted the introduction of electronic ballasts.

## Lamp Operation

Fluorescent ballasts operate fluorescent lamps in the following ways:

**Preheat**: Lamp electrodes are heated prior to the application of a high starting voltage that initiates the arc discharge. The starting electrode voltage is applied through a "starter," a thermal switch that, when it opens, applies the high starting voltage across the electrodes. No power is applied to heat the cathodes during operation. Lamps of less than 30 watts are usually operated preheat. Preheat operation is characterized by lamp "flickering" when starting.

**Rapid-start**: Lamp electrodes are heated prior to and during operation by two low-voltage windings on the ballast. Rapid-start operation is characterized by smooth starting and long lamp life.

**Instant-start**: Lamp electrodes are not heated. Instant-start ballasts provide a high open-circuit voltage across the unheated electrodes to initiate the arc discharge. Instant-start operation is more efficient than rapid start, but as in preheat, lamp life is shorter. Eight-foot "slimline" lamps are operated in the instant-start mode.

## Magnetic Ballasts

All ballasts have electrical and magnetic losses. These internal losses and the ability of the ballast to operate the lamps efficiently contribute to the overall efficiency of a ballast. The efficiency of magnetic ballasts have been improved by using low-loss magnetic material and copper windings, resulting in lower internal losses.

## Ballast Efficacy Factor (BEF) and Ballast Factor (BF)

Ballast efficiency is regulated by the Federal Appliance Standard that prohibits the manufacture of ballasts that do not meet a minimum BEF. BEF for a specific lamp/ballast combination is calculated by dividing the percent rated light output by the measured input power in watts. The percent rated light output is found by multiplying the ballast factor by 100%. BEF can be confusing because it is meant to compare the performance of ballasts on a specific lamp and thus is not of particular value in evaluating efficiency opportunities.

BF is the ratio of the lamp lumen output on a commercial ballast to the lamp's rated light output (lamp catalog rating). Fluorescent lamps are rated in lab tests using a lossless, perfect reactor ballast, so ballast factor is required to derate the catalog rating of lamps.

## Electronic Ballasts

The efficiency of fluorescent ballasts can be improved beyond the use of energy-efficient magnetic ballasts. When fluorescent lamps are operated by an electronic ballast at high frequency, they convert the input power to light output more efficiently. The lumens per watt (LPW) of the lamp/electronic ballast combination is increased, which means either producing more light for the same power or producing the same light with lower power. The total system input watts can actually be less than the total of the lamp wattages. In fact, when a 32-watt T8 lamp is operated at high frequency, it consumes only about 28 watts, making the system input watts about 60 watts. The actual watts input is controlled by the ballast factor, which can range from 47 to 130%. Generally, the lower the ballast factor, the lower the input watts. Proper selection of all electronic ballast parameters are important for successful lighting applications.

## Tandem Wiring

Multilamp fluorescent ballasts, controlling up to four lamps, may be wired to operate lamps outside the fixture where they are mounted. When the fixtures are not in continuous rows, lead length limitations and National Electric Code requirements for protection of the wiring must be considered. Care needs to be exercised to ensure that the correct set of leads is used to wire lamps in the remote luminaire. Some ballast manufacturers provide special labeled ballasts for tandem applications. Inboard/outboard switching of 3- or 4-lamp luminaires is accomplished using multilamp ballasts in each luminaire pair and tandem-wiring the two luminaires.

## Fact Sheet 5: Luminaires

Unlike fixtures, luminaires are complete lighting units, consisting of (1) lamps and ballasts to convert power to light, (2) lamp holders, (3) an optical system, (4) a means for connecting to power, and (5) a housing. Luminaires distribute, direct, and diffuse light.

### Optics

The optical system includes the lamp cavity and diffusing media and includes one or more of the following components: reflectors, refractors, diffusers, lenses, baffles, or louvers.

**Reflectors** redirect light by using the principle of reflection. Retrofit reflectors upgrade the performance of fluorescent luminaires by increasing the interior reflectance.

A **refractor** is a component that redirects light by refraction.

**Lenses** are made of transparent or translucent materials that have a prismatic configuration on the bottom surface to alter the directional characteristics of light passing through it. Polystyrene lenses are the least expensive, but such lenses turn brownish-yellow due to the ultraviolet (UV) radiation from the lamps. Lenses made from acrylic are light-stabilized and do not discolor.

Translucent sheets of milky-white plastic, called **diffusers**, shield the lamps and scatter light uniformly in all directions below the ceiling plane, reducing source brightness.

**Baffles** are one-way louvers, made of metal or plastic.

**Louvers** consist of baffles arranged in a geometric pattern that control the light distribution and shield the source from view at certain angles. Louvers range from small-cell plastic louvers to large-cell aluminum louvers. Parabolic luminaires use large-cell louvers formed in a parabolic shape. The resultant light distribution reduces glare, controls light output, and has high aesthetic appeal. Paracube louvers are small metallic-coated plastic squares; 1/2-inch squares are the most common size. Small-cell louvers have high visual comfort probability (VCP) but sacrifice luminaire efficiency and spacing criteria.

The highly polished surfaces used in some louvers and reflectors are called specular finishes. Semi-specular finishes reflect some of the light directionally, with some amount of diffusion.

### Types of Luminaires

Interior luminaires can be direct (downward), indirect (upward), or direct-indirect.

**Direct luminaires** can be open or shielded. Shielded luminaires use lenses, baffles, or louvers. Downlights are direct ceiling luminaires that direct all the light downward. They can be recessed, surface-mounted, or suspended. Recessed units include HID downlights and a wide range of fluorescent troffers. Recessed troffers are luminaires mounted within the ceiling so that the bottom surface is in the same plane as the ceiling. Some troffers are designed to lay on the tee-bar flanges of a suspended grid ceiling system. Surface-mounted troffers are attached directly on the ceiling and project below the ceiling by their depth. Pendant-mounted units such as direct-indirect luminaires are supported from the ceiling structure by stems, pendants, or aircraft cable, that position the luminaire below the ceiling. **Air-handling** is a term applied to a recessed luminaire that supplies or returns comfort air by means of slots. A static luminaire does not handle comfort air.

**Indirect luminaires** radiate light up to a reflecting ceiling. Indirect lighting systems usually use luminaires suspended from the ceiling by aircraft cables. Cove lighting, wall-mounted units and furniture-top units are also considered indirect lighting. Lighting offices with indirect luminaires results in higher, more uniform brightness on both ceilings and walls, and glare on computer screens is reduced significantly. Shadows are reduced from partitions and under-shelf cabinets. The success of indirect systems depends on maintaining a high ceiling reflectance. Recent designs use T8 lamps (or T5) and electronic ballasts in a low-profile housing with lenses and special reflectors to achieve high luminaire efficiency and a broad distribution while allowing mounting close to the ceiling.

**Direct-indirect** luminaires combine the efficiency and high coefficient of utilization (CU) of direct luminaires with the uniformity and glare control of indirect luminaires. Some office and school systems are designed for an equal balance of direct downlight and indirect uplight.

### Luminaire Light Loss Factors

Luminaire surface depreciation is the nonrecoverable light loss factor (LLF) that describes how light is lost as the reflecting surfaces degrade over time. Luminaire dirt depreciation (LDD) is the recoverable light loss factor that describes how much light is lost from the initial illuminance provided by clean, new luminaires to the reduced illuminance that will be provided after dirt collects on the reflecting surfaces at the time at which it is expected that cleaning will be done. Another light loss factor that causes light loss from luminaires is lamp lumen depreciation (see Fact Sheet 1).

## Fact Sheet 6: Luminaire Identification

**Luminaire** is the term lighting professionals use to refer to what are commonly called fixtures. Identifying different luminaire types is a matter of learning the physical shapes, nominal sizes, and names of the different luminaire families (see also Fact Sheet 5).

### Classification of Luminaires

In this handbook we classify luminaires in three major categories:

1. interior—fluorescent,
2. interior—industrial, and
3. exterior.

### Interior — Fluorescent

Figure 8.6 shows 13 luminaires that may be found in interior spaces. Decorative chandeliers and sconces, commonly lamped with incandescent sources and found in hospitality spaces, have not been shown.

Single-lamp and two-lamp strip lights range from 2 feet to 8 feet and can be found in indirect cove and cornice lighting and in direct luminous ceiling applications. Sometimes a lens cover is used, and in some industrial applications it may be sealed and gasketed. The two-lamp open fixture is found in its 8-foot size in many industrial applications, usually in continuous rows and lamped either with slimline (F96T12) or HO lamps. The pendant-mounted direct or direct-indirect luminaire is found in schools and older offices, usually in its 4- or 8-foot size and usually in continuous rows lamped either with slimline or HO lamps.

The suspended luminaire pictured in Figure 8.6 is suspended from a ceiling by aircraft cables and has a power feed at the left end. Three types of recessed troffers are shown in Figure 8.6.

When the auditor encounters luminaires with louvers, it is helpful to note the number of cells. Note that the louvered recessed troffer on the left in Figure 8.6 has a 3 × 6 cell array, totaling 18 cells. The troffer on the right-hand side is equipped with a 9-cell louver. These luminaires will usually be equipped with U-tubes.

If an older, ceiling surface-mounted round direct luminaire like the one in Figure 8.6 is encountered, the auditor should take the diffuser/cover off to discover if the unit is lamped with incandescent, circleline fluorescent, or compact fluorescent lamps.

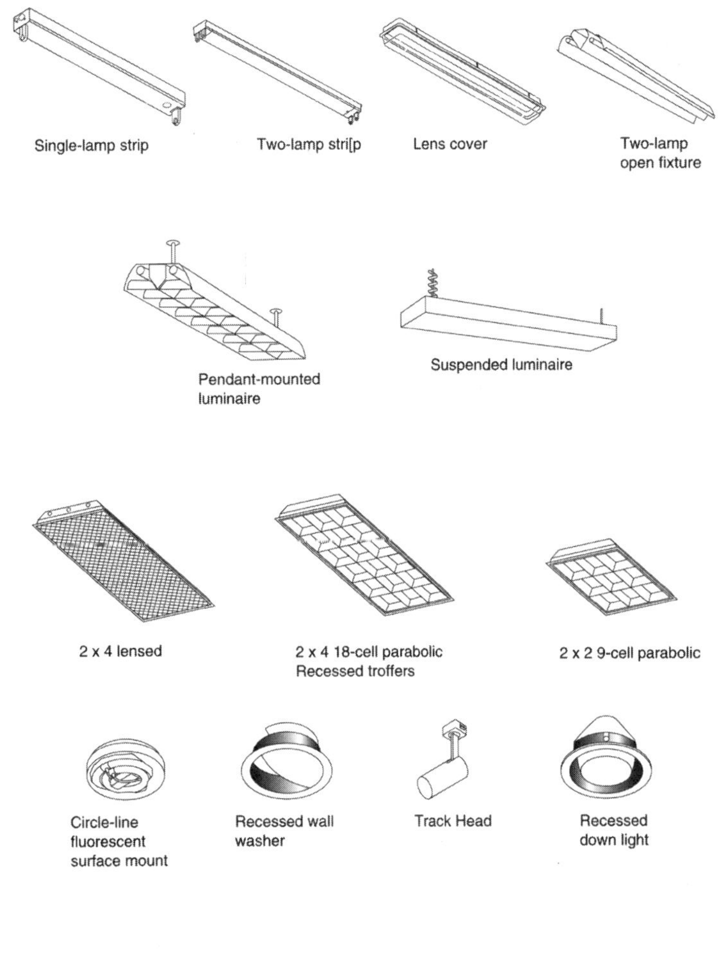


**FIGURE 8.6** A variety nof luminaires that may be found in interior spaces (see text for details about the luminaires pictured).

A recessed wall washer can be found lamped with incandescent A lamps, PAR lamps, or CFLs. Wall washers are usually located in a ceiling, mounted a few feet from a wall, and should function to light that wall.

Track lighting is easy to spot, and it is usually equipped with incandescent or tungsten-halogen bulbs. Downlights come in several aperature opening diameters and are equipped with incandescent, CFL, or low-wattage metal-halide lamps.

### Interior — Industrial

Figure 8.7 shows six industrial-type luminaires that are usually found equipped with HID bulbs.

### Exterior

Figure 8.8 shows nine varieties of luminaires used in outdoor (exterior) applications. These exterior luminaires are usually found equipped with HID lamps, such as metal-halide or high-pressure sodium.

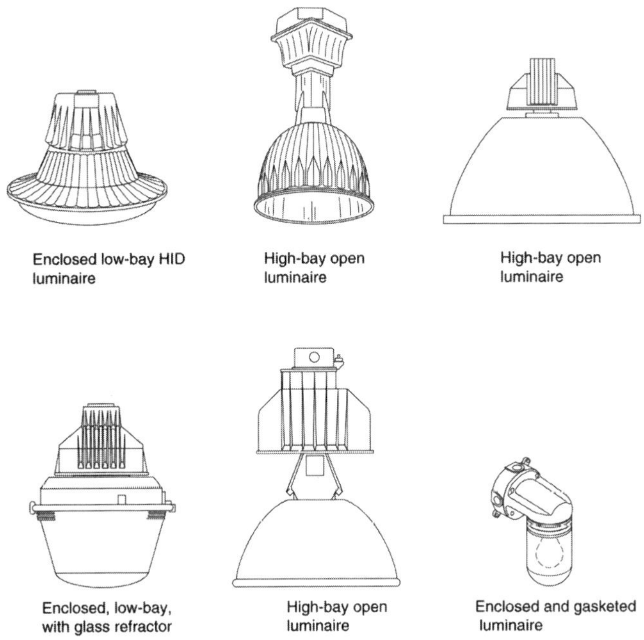


**FIGURE 8.7** Some industrial -type luminaires. The high-bay open luminaires shown have ballast units mounted above their reflector or refractor assemblies. (Someimes the ballast unit is mounted remotely.) The low-bay luminaires shown are enclosed and are usually found at mounting heights less than 20 feet. The glass refractor on one of the units pictured controls light distribution. The small enclosed luminaire pictured may be gasketed and may be mounted in a severe environment, if it has been rated for the particular environment. Lamps used in such environments may be incandescent, CFL, or HID.

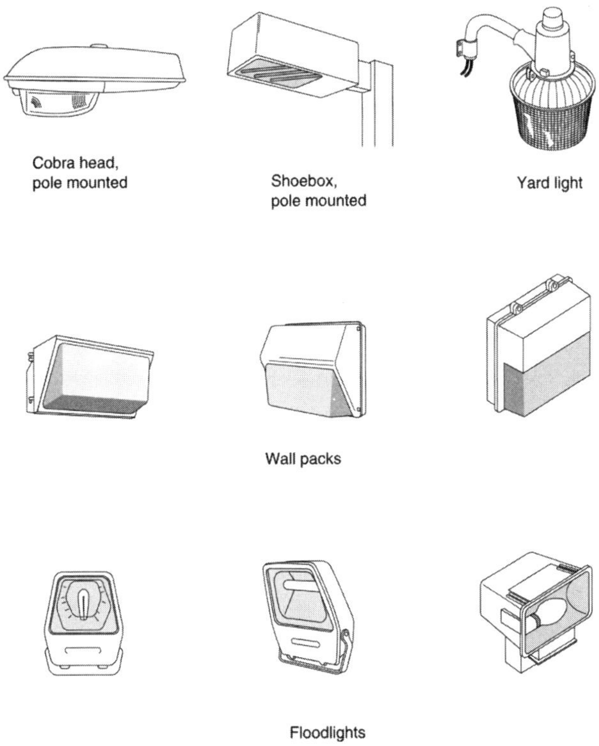


**FIGURE 8.8** A variety of exterior luminaires. The pole-mopunted "cobra head" is used in roadway or walkway lighting. A "shoebox" luminaire may be found in parking lots and walkway lighting. Wall packs are usually used to light areas near the wall where they are mounted. Floodlights may be found lighting building facades and parking lots. Vertically mounted lamps and horizontally mounted lamps produce different beam patterns. A compact HID floodlight may be used to highlight small areas of a building.

# Fact Sheet 7: Lighting Systems

Lighting systems can be classified by light source, by location (interior/exterior), and by light distribution.

## Incandescent Systems

This category includes recessed downlights, using R, ER, PAR, or A lamps; track lighting using R or PAR lamps; wall sconces; chandeliers with decorative sources; and table lamps using standard A lamps. It also includes luminaires that are designed to use quartz or tungsten-halogen sources.

## Fluorescent Direct Systems

Recessed troffers are recessed above a grid ceiling system, consisting of T-bars, and ceiling tiles. Troffers are 1 foot × 4 feet (1×4), 2 feet × 2 feet (2×2), or, most commonly, 2 feet × 4 feet (2×4), although there are also some with 20-inch widths. Smaller units, usually found in corridors, may be as small as 1 foot × 1 foot. The center-to-center spacing in the grid is set by the grid openings, with the maximum spacing set by the spacing criteria of the luminaire (see Fact Sheet 5). Fact Sheet 13 shows the calculation for spacing using the spacing criteria (SC). Common layouts are shown in Fact Sheet 8. They may be either air-handling—providing (supply) or returning (return) comfort air—or static (non-air-handling). For plaster ceilings or grid ceilings with no space above, fixtures are usually mounted to the ceiling and are classified as surface-mounted. Fixtures are suspended below the ceiling from pendants or aircraft cables for direct/indirect and indirect luminaires. Lamps used may be T12 (1 1/2"D), T10 (1 1/4"D), T8 (1"D), T5 (BIAX or Dulux L), or smaller compact fluorescents. The lamps may be vertical T5 (5/8"D).

## High-Intensity Discharge (HID) Direct Systems

Usually found in high ceiling applications (above 10 to 12 feet), HID systems employ metal-halide or mercury lamps. Recessed fixtures include downlights, using lower-wattage metal-halide or mercury lamps. There are 2×2 fixtures equipped with internal refractors and an external lens (to control glare) that employ higher-wattage (usually up to 400 watts) metal-halide lamps.

## Fluorescent Indirect Systems

Indirect systems use luminaires suspended (by pendants or aircraft cables) from the ceiling that radiate light up to a reflecting ceiling. Fluorescent indirect may also be furniture-mounted and use the lamps listed previously under "Fluorescent Direct Systems."

## High-Intensity Discharge (HID) Indirect Systems

Popular in the seventies, these systems use large hockey-puck–shaped fixtures, usually suspended from pendants from the ceiling or are mounted on the tops of office furniture or in kiosks. They use metal-halide or mercury sources up to 400 watts.

## Luminous/Louvered Ceilings

These systems, popular in the sixties, employ a translucent media or "egg crate" louvers to cover a large section or an entire ceiling. The media may be glass or plastic and the louvers may be metal or plastic. They are usually lighted by fluorescent strip lights, with 4' T12, 8' Slimline, HO, or VHO lamps. The best uniformity requires closely spaced rows of strips that result in power densities as high as 5 W/sf, making them notoriously inefficient.

## Task/Ambient Lighting

Found in office lighting with heavy computer use, these systems use a combination of an ambient system to provide a lower light level (on the order of 20 to 30 FC), usually from a ceiling direct system, supplemented with task lighting provided by furniture-mounted fixtures or separate desk lamps.

## Industrial Lighting Systems

For the past several decades, industrial systems have relied on incandescent, mercury vapor, metal-halide, high-pressure sodium, and fluorescent sources to light manufacturing spaces. Incandescent and mercury vapor lamps are obsolete sources for general-purpose industrial lighting today. Linear fluorescent fixtures, equipped with 4' T12 (or more recently T8), 8' Slimline, HO, or VHO lamps are generally used at lower mounting heights. These may be simply open strip lights in continuous rows, sometimes equipped with reflectors to redirect the light. The HID sources (high-pressure sodium, mercury, or metal-halide) are preferred for higher mounting heights and are found in high-bay luminaires, mounted at about 20 to 25 feet, or in low-bay luminaires, mounted at less than 20 feet.

## Emergency Lighting Systems

Emergency lighting equipment is not usually considered part of building lighting systems. The selection and location of emergency units are determined by the National Fire Protection Association (NFPA) Life Safety Code (101). Unit equipment consists of battery powered incandescent floodlight "heads" that are activated upon loss of building power. EXIT lights are continuously operated to designate the building egress routes. Recent advances in exit light retrofit equipment and the long burn hours required of exit lights make them good candidates for upgrading. Emergency lighting equipment is often backed up by an emergency generator system.

## Exterior Lighting Systems

These systems include street, roadway, parking lot, walkway and building facade lighting. Parking lot, walkway, and street lighting systems are usually mounted in a fixed arrangement. When floodlighting is used to light parking lots and building facades, the units are aimed at a predetermined point. There are seven NEMA beam spread classifications for floodlighting. Luminaires have been designed for parking lot applications to eliminate the direct glare of floodlights in drivers' eyes. These units are mounted in or near horizontal position on poles at mounting heights below 45 feet, and they throw a rectangular beam pattern. Large parking lots are sometimes lighted with "high mast" equipment using pole heights of 80 to 150 feet and arrays consisting of 10 to 20 luminaires using up to 1000-watt HID lamps, usually metal-halide or high-pressure sodium. Lowering devices are used to lower the arrays for relamping and other maintenance.

Street lighting employs equipment that is specifically designed for lighting streets. Most street lighting overhangs the roadway and uses asymmetric distributions to project light along and across the roadway, with just a small amount of light falling on the narrow strip of roadway behind the fixture. There are five distribution classifications, based on light distribution and on the distance they project light along the street (long, medium, or short throw).

# Fact Sheet 8: Interior Lighting Layouts and Ceiling Plan Guides

Auditors can learn much from examining ceiling layouts while performing building lighting surveys and audits. Interior lighting layouts are usually described in terms of "rows" and "columns" of luminaires as shown in Figure 8.9.

## Grid Ceiling Layouts

Grid-type suspended ceilings are the ceiling type most frequently found in commercial spaces. Regular patterns of rows and columns of fluorescent 2×4 luminaires are most common. This fact sheet shows a common grid (suspended) ceiling pattern of 8 feet by 8 feet. Grid ceiling patterns are named after the distances the luminaires are located from each other. Hence, the 8×8 pattern shown in Figure 8.10 measures 8 feet from the centerline of one luminaire to the centerline of the adjacent luminaire in the direction perpendicular to the lamps. In the direction parallel to the lamps, the distance is 8 feet, again measured from the centerline of the adjacent luminaire.

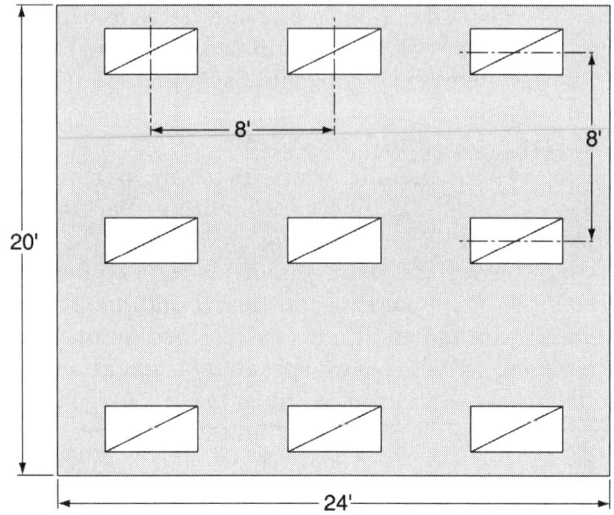


**FIGURE 8.10** Grid ceiling pattern, 8 × 8 centers.

## Luminaire Spacing

Luminaires have a maximum spacing limit set by the spacing criteria (SC) of the luminaire. The photometric distribution resulting from a particular luminaire design determines the spacing criteria. Spacing criteria is given as two values for nonsymmetrical luminaires—one value for each direction (parallel to the lamps and perpendicular to the lamps). The center-to-center spacing in a grid is set by the grid openings, with the maximum spacing set by the spacing criteria of the luminaire. Spacing criteria is detailed in Fact Sheet 5, and Fact Sheet 13 shows the spacing calculation.

Luminaire layouts should be recorded by the auditor. Nonstandard layouts should be sketched. This layout and spacing data is valuable for upgrades that involve modification of the luminaire light distribution, such as reflectors, and for relighting projects that change the type of luminaire. Relighting projects use this information to compare and describe the new luminance distribution. The layout information will also be useful for lighting power distribution considerations.

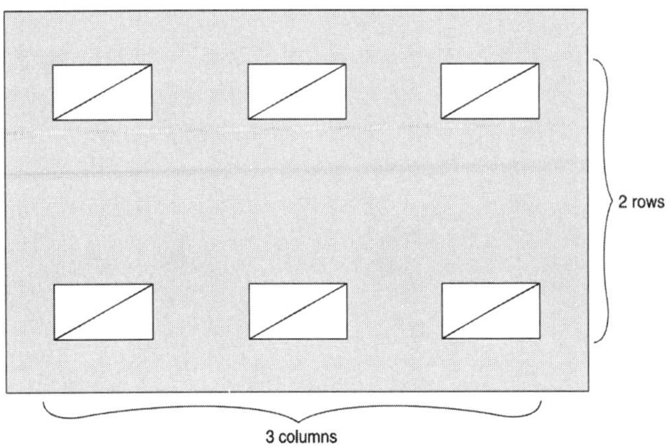


**FIGURE 8.9** Interior lighting layouts are usually described in terms of rows and columns of luminaires.

## Estimating Fixture Count from Fixture Layout

The fixture count in rooms can be estimated from the layout. For example, for a grid ceiling with 2×4 luminaires in an 8×8 pattern, each luminaire covers 64 square feet. In a 700-foot-square room with an 8×8 pattern, the fixture estimate would be 10.

## Estimating Power Density (PD) and Illuminance (FC)

Since grid ceilings only allow fixtures to be placed at regular intervals in the grid, tables can be used to estimate the power density and illuminance (light level) resulting from each pattern and for particular combinations of equipment. Table 8.2 shows this for 2×4 troffers ranging from the least efficient, T12 lamps and standard core and coil magnetic ballasts, to the most efficient, T8 lamps with electronic ballasts.

For a given pattern, such as the 8×8 pattern shown in Table 8.2, power density values range from 1 W/sf to nearly 4 W/sf, depending on the lamp and ballast combination. Note that the more efficient the lighting equipment, the lower the power density. Values of maintained illuminance range from 48 to 112 footcandles, based on a light loss factor (LLF) of 0.75 (see Fact Sheets 5, 13, and 14).

If the auditor first identifies the type and quantity of lamps and the type of ballasts in the luminaires of a facility, then it is simply a matter of looking up values in a table to determine the power density and the initial illuminance for any pattern.

**TABLE 8.2** Power Density and Light Level Values for an 8 × 8 Pattern

| Lamp and Ballast Type | Lamps/Fixture | Watts/Fixture | Power Density (W/sf) | Maintained Light Level, in FC (LLF = 0.75) |
|---|---|---|---|---|
| T8/LBF-ELB | 2 | 51 | 1.0 | 48 |
| T12RW/ESB | 2 | 72 | 1.4 | 52 |
| T1240/STD | 2 | 95 | 1.8 | 60 |
| T8/LBF-ELB | 3 | 91 | 1.7 | 70 |
| T12RW/ESB | 3 | 107 | 2.0 | 76 |
| T1240/STD | 3 | 143 | 2.7 | 88 |
| T8/LBF-ELB | 4 | 109 | 2.0 | 88 |
| T12RW/ESB | 4 | 138 | 2.6 | 96 |
| T1240/STD | 4 | 190 | 3.6 | 112 |

*LBF = low ballast factor (0.8); ELB = electronic ballast; ESB = energy-saving ballast; STD = older coil and core ballast; RW = reduced wattage (34-watt Watt-Miser, Super-Saver, Econo-Watt)*

## Fact Sheet 9: Improving Lighting System Performance

Lighting can account for 30 to 50% of the total electricity used in commercial buildings. An estimated one-third to one-half of this electrical use can be reduced by improving lighting system performance. Savings can be accomplished in two ways: by reducing the lighting load (watts) by means of system efficiency improvements and by reducing the time the systems are operated, through the use of controls.

### Building Systems Interaction

Lighting systems are an integral part of commercial building systems. There is an interdependent relationship between lighting, building envelope, and HVAC systems. Electric lighting contributes to the internal heat gain of a building and must be accounted for when considering the capacity of an HVAC system. If an efficient lighting system that generates less heat is used, the air-conditioning load will be less. This reduction increases energy savings over the direct savings due to the decrease in lighting load. Heating requirements will increase, but for most commercial buildings there is a net financial savings because the heating fuel cost is lower than the cooling cost.

### Measures of Performance

Power density is the power allowance for a building, measured in watts per square foot (W/sf). Power density is a good figure of merit for the efficiency of lighting systems. In current practice, new, efficient office lighting systems are measuring down to 1.0 W/sf (without daylight contribution). Systems that calculate above this value may be considered inefficient and candidates for upgrade or replacement. Efficacy in lumens per watt (LPW) is a measure of performance for lamps or lamp and ballast combinations. Fluorescent lamp efficacies presently range up to 100 LPW.

### Luminaire Performance

Luminaire performance is a function of luminaire efficiency and the lighting application. Luminaire efficiency is expressed as a percentage—the ratio of lumens emitted by a luminaire to lumens emitted by the lamp(s) alone.

$$\text{Luminaire efficiency} = \frac{\text{lumens from luminaire}}{\text{lumens from lamps only}} \tag{8.1}$$

There are three separate components of luminaire efficiency: electrical, optical, and thermal. Electrical efficiency is defined by the power-to-light conversion and depends on the lamp and ballast combination. Optical efficiency is determined by the luminaire optical components (see Fact Sheet 5) that define how the light gets out of the luminaire. Thermal efficiency depends on the interaction of the lamps and their thermal sensitivity (see Fact Sheet 11).

The coefficient of utilization (CU) defines how a given luminaire will perform in a particular room geometry with certain room surface reflectances. CU includes luminaire efficiency and the intensity distribution. The CU is expressed as a percentage of the lumens produced by the lamps that reach the work plane. CU tables are found in luminaire manufacturers' catalogs.

### Upgrading Luminaire Performance

Luminaire performance can be improved in several ways: with the use of specular reflectors, with a lens or louver upgrade, by regular cleaning of the luminaires, or by replacement with new, higher efficiency luminaires. Reflectors can improve 2×4 troffer efficiency by up to 17%—and by even more if the troffer interior is deteriorated. For maximum performance, the lamp centers need to be relocated and a custom reflector designed for the particular luminaire. Inexpensive styrene lenses that have become discolored can block a high percentage of light from the luminaire. Replacement with a high quality acrylic lens or a special high-performance lens can also increase performance. There are also retrofit louver kits to convert lensed luminaires. Maintenance is a critical factor in luminaire performance. The accumulation of dirt on lamp and optical component surfaces and lack of cleaning can cause the output of luminaires to decrease by as much as 40%. Scheduled maintenance can reduce both initial and operating costs (see Fact Sheet 14).

### Deciding Whether to Upgrade or to Relight with New Luminaires

The decision to upgrade luminaire performance or relight depends on several considerations, including the age and condition of the luminaires, how easily they can be retrofit, and the economic payback. If lamp sockets have to be relocated or replaced, it is often less expensive to purchase new luminaires. If the lighting system can be made more efficient by a change to a different light source, and if there is the opportunity to use a different quantity of luminaires or a different layout, then relighting with new luminaires should be given strong consideration. Fewer luminaires can be used if regular planned maintenance is assured. Another consideration for relighting is that there are more electrical contractors that can install new lighting than there are that can perform extensive luminaire upgrades. The lighting efficiency rating needs to be considered for new luminaires.

## Luminaire Efficacy Ratings (LERs) for Fluorescent Luminaires

The National Electrical Manufacturers Association (NEMA) has developed a procedure for determining the luminaire efficacy rating (LER) for fluorescent luminaires. This procedure is described in NEMA standards publication No. LE 5. The LER is specified as a minimum value for each luminaire type and classification. There are currently five high volume luminaire types and classifications listed in the LE 5-1993. The LER of each luminaire type is calculated by the testing lab for a given luminaire by the following formula:

$$\text{LER} = (\text{EFF} \times \text{TLL} \times \text{BF})/\text{Watts Input}$$

where:

LER = Luminaire efficacy rating, expressed in rated lumens per watt

EFF = Luminaire efficiency, obtained from a standard photometric test performed according to IESNA LM 41 and expressed as a two-place decimal

TLL = Total lamp lumens

BF = Ballast factor of test ballast or the average BF of test ballasts used in the photometric test, shown as a two-place decimal

Watts Input = Total wattage of the luminaire as measured during the photometric test, shown as a whole number

## Other System Efficiency Improvements

There are other system efficiency improvements that can be made in addition to upgrading luminaire efficiency. For example, when single-lamp ballasts are encountered, replacement with multilamp electronic ballasts will always increase efficiency, since multilamp units are more efficient than single-lamp units. In fluorescent cove or coffer lighting, luminous ceilings, and other systems that employ continuous rows of fixtures, multilamp ballasts that are tandem-wired are usually cost-effective (see Fact Sheet 4).

## Controls

Lighting controls will ensure that electrical energy (kWh) is not consumed unnecessarily. Lighting controls can save lighting energy by controlling the time (hours) lighting is used and by controlling the lighting load (kW), usually through dimming. Occupancy sensors, timers, photosensors, dimming equipment, and centralized control systems are available to determine when lighting should be provided, regulate the amount of light, and control the task lighting of a particular area.

## Fact Sheet 10: The Effects of Overvoltage and Undervoltage on Lighting Systems

Fluorescent lighting systems are affected by the voltage applied to the ballast. The effects of overvoltage and undervoltage are observed in (1) lamp starting, (2) light output, and (3) lamp life. Overvoltage can affect ballast life.

### Normal Fluorescent Operating Voltage

The normal operating range of input voltage to the ballast is -7.5% to +5% of the nominal voltage. Some electronic ballast designs incorporate voltage regulation and will accept wider ranges of input voltages without affecting lamp starting, light output, and lamp life.

### Lamp Starting

Low line voltage—voltages below -7.5% of the nominal voltage—can result in unreliable starting. Starting problems, in turn, can cause reduced lamp life. The ballast must provide a high enough starting voltage to initiate the discharge arc, and if line voltage is low, the starting voltage may also be low.

### Light Output

Low line voltage reduces lamp current, which reduces light output and extends lamp life, provided that sufficient cathode heating is provided by the rapid-start ballast. When the input voltage is lower than normal and cannot be changed, a voltage factor is included in lighting calculations to account for the loss of light output. High line voltage increases light output and decreases lamp life. The effect of lamp current on lamp life is shown in Figure 8.11.

### Lamp Life

Higher-than-normal line voltage will increase the starting voltage which, in turn, will result in excessive cathode erosion, reducing lamp life and causing darkening of lamp ends. In rapid-start and preheat systems, too high an input voltage results in high cathode temperature, which reduces lamp life. If the input voltage is too low, it results in low cathode temperature, which causes cathode sputtering and thus reduces lamp life.

### Ballast Life

Fluorescent, magnetic ballast life is affected by overvoltage. Higher-than-normal line voltage will increase ballast heating, reducing ballast life. A 5% increase in line voltage (above the normal 5% limit) will increase the ballast temperature 10 degrees F and result in a 50% reduction in life.

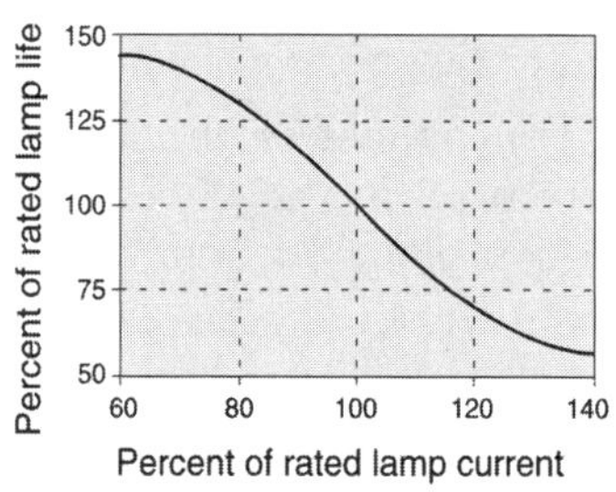


**FIGURE 8.11** The effect of lamp current on lamp life.

## Fact Sheet 11: The Effect of Ambient Temperature on Lighting Systems

Fluorescent lamps are temperature-sensitive, and their light output is affected by ambient temperature. This effect is shown in Figure 8.12. Ambient temperature is the temperature around the lamp within the luminaire. The coldest point on the lamp bulb wall of a fluorescent lamp that defines the point where excess mercury condenses is called the minimum bulb wall temperature (MBWT).

Light output is a function of mercury vapor pressure, which is directly related to MBWT. The optimal MBWT typically occurs at 25°C, which is the temperature used in standard test conditions for lamps.

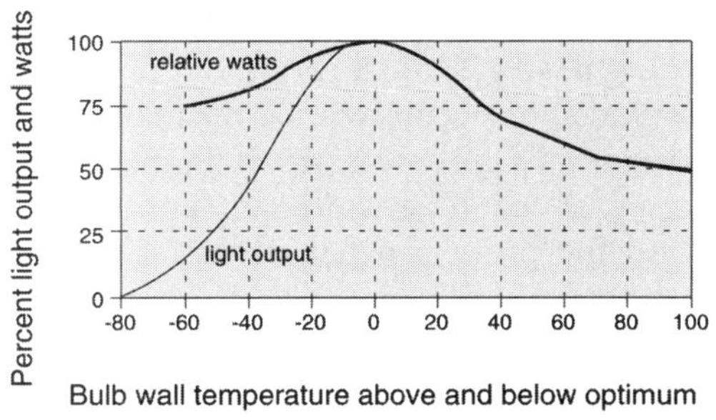


**FIGURE 8.12** The light output of fluorescent lamps is effected by ambient temperature.

### Normal Fluorescent Operating Temperature

The normal operating range for rapid-start systems is 50 to 110 degrees F. For instant-start operation the ambient temperature must be greater than 0 degrees F. Reduced-wattage lamps—Super-Saver, Econo-Watt, and Watt-Miser—require an ambient temperature above 60 degrees F to operate properly.

### Out-of-Tolerance Operating Temperature

If the ambient temperature is too high, it can cause lamps to have starting problems. As shown in Figure 8.12, the light output will be reduced at elevated temperatures. Light output is more severely affected if the ambient temperature is below normal and starting becomes more difficult. In addition, the system efficacy, measured in lumens per watt (LPW), will be reduced, since the light output drops off faster than the input watts.

### High-Intensity Discharge (HID) Operating Temperature

The light output of double-envelope HID lamps is not affected by ambient temperature, and normal operation can be observed down to -20 degrees F.

## Fact Sheet 12: Effects of Switching on Lamp Life

When fluorescent lamps were first manufactured in the 1940s, electricity costs were low. These early lamps burned out more quickly if switched on and off daily. For this reason, companies left their fluorescent systems on day and night. At that time, because of the extremely low electric rates, the electricity actually cost less than the labor and materials needed for lamp replacement. However, with technology advancements and increased electricity costs, the conventional wisdom of operating fluorescent systems for long periods of time can no longer be supported.

### Myths and Realities

There are two myths that control how some companies still operate their fluorescent lighting systems.

**Myth 1**: Continuously operating fluorescent lights is cheaper than turning them off for brief periods.

**Reality 1**: Turning off fluorescent lights saves energy, extends overall lamp life, and reduces replacement costs.

**Explanation 1**: Many users believe that there is a significantly higher electricity consumption during the starting of a fluorescent lamp than when the lamp is operating for long periods. However, according to tests performed at the U.S. Naval Civil Engineering Laboratory, the initial starting surge lasts only 1/120 second and the entire starting current for a two-lamp rapid-start luminaire lasts for less that one second before stabilizing.

**Example 1**: Turning off one F40 fluorescent lamp for 1/2 hour each day saves $4.00 at $0.08/kWh in energy over the life of the lamp.

**Myth 2**: Lights shouldn't be turned off because it shortens lamp life and increases maintenance costs.

**Reality 2**: Fluorescent lamps will run more hours if operated continuously, but they will last for many more years if they are turned off when not in use.

**Explanation 2**: Although the average rated life of fluorescent lamps is shortened by switching, calendar life is lengthened. Calendar life, sometimes called effective lamp life, is the time between lamp changes and includes the time the lamp is off.

**Example**: Standard F40 rapid-start lamps operated continuously result in a rated lamp life of 34,000 hours, which is equivalent to an effective or calendar life of *3.9 years*. Turning off F40 lamps for 12 hours each day will decrease the average rated lamp life to 30,000 hours, but the effective lamp life is extended to *6.8 years*!

### Lower Lighting Costs

Lamp replacement costs will be reduced by turning off lighting systems and electric bills will be reduced. The money saved by turning off lights is usually more than the price of a new lamp.

### Effects of Frequent Switching on Fluorescent Lamps

Table 8.3 shows fluorescent lamp life for various on and off times. Lamp life must be balanced against energy use and the average number of times the lamp is switched on and off during the day, as well as maintenance and lamp costs.

Since electricity accounts for the majority of lighting costs, it is usually more cost-effective to switch lights off when they are not needed. If a room is unoccupied for as little as 5 minutes every hour, it is more economical to turn the lights off while the room is vacant.

In cases where lights are switched on and off more frequently, it can be more economical to leave lights on for a certain period of time. For example, in a restroom where lights are turned on and off every 5 minutes, it may be cost-effective to leave them on. When lamp life is weighed against the cost of energy used, the most economical solution is to leave the

**TABLE** 8.3 Effects of Frequent Switching on Fluorescent Lamps

| On Time (hours per cycle) | Off Time (hours per cycle) | Total Lamp Life (hours) | Total Lamp Life (years) | Energy Cost (% of Base) |
|---|---|---|---|---|
| 9 | 0 | 27,000 | 12.0 | Base |
| 3 | 0.5 | 20,000 | 10.0 | 89% |
| 2 | 0.5 | 17,000 | 9.1 | 83% |
| 1 | 0.25 | 13,000 | 8.7 | 67% |
| 0.5 | 0.25 | 8,000 | 5.3 | 67% |
| 0.25 | 0.25 | 6,000 | 5.3 | 50% |

*Source: Electric Power Research Institute*

lights on for 10 to 20 minutes before turning them off. This strategy can be implemented by setting the delay time adjustment of an occupancy sensor for 10 to 20 minutes.

The use of rapid-start ballasts are recommended for frequent switching applications. Using instant-start ballasts for frequent switching applications can result in disappointing lamp life.

## Incandescent and High-Intensity Discharge (HID) Lamps

Turning off incandescent lamps in unoccupied rooms can save considerable energy, especially where higher-wattage bulbs are used. Frequent switching only slightly affects incandescent lamp life. Because of long re-strike times, HID lamps should be turned off when not needed for 20 minutes or more.

## Fact Sheet 13: Calculations

The lighting auditor needs to know several types of calculations.

### Lighting Power Density (PD)

Sometimes referred to as the unit power density (UPD), lighting power density is the power allowance for a building, measured in watts per square foot as shown in Equation 8.3:

$$\text{PD} = \frac{\text{total watts of lighting system(s) in audited space(s)}}{\text{area of audited space(s) (square feet)}}$$

Example:

$$\text{PD} = \frac{1050 \text{ watts}}{525 \text{ square feet}} = 2.0 \text{ W/sf}$$

Power density is a good figure of merit for the efficiency of lighting systems. In current practice (1997), new, efficient office lighting systems are measuring down to 1.0 W/sf (without daylight contribution). Systems that calculate above this value may be considered inefficient and candidates for upgrade or relighting.

The power density (PD) can be used by the auditor as a retrofit screening measure. If a particular lighting retrofit is calculated to be cost-effective for spaces of a certain PD, then all similar spaces measuring higher than that PD can also be upgraded cost-effectively.

### Average Light Level

Minimum and maximum readings are located and an average illuminance (in footcandles) is calculated by adding the two readings and dividing by 2 as shown in Equation 8.4.

$$\text{FC (avg)} = \frac{\text{FC (max)} + \text{FC (min)}}{2} \qquad (8.4)$$

Example:

$$\text{FC (avg)} = \frac{60 + 30}{2} = 45 \text{ FC}$$

### Zonal-Cavity Method

Used to calculate average illuminance, in footcandles or lux, this method for simple room shapes requires that the user look up a coefficient of utilization (CU) from a CU table in the luminaire manufacturer's catalog. Prior to calculating the initial illuminance, a room cavity ratio (RCR) calculation is performed, to be used when looking up the CU. Equation 8.5 shows the calculation for initial illuminance, in footcandles. If lux is desired, the units of the space measurements are changed to meters.

$$\text{FC (initial)} = \frac{\text{\# of fixtures} \times \text{\# of lamps/fixture} \times \text{lumens/lamp} \times \text{CU}}{\text{length} \times \text{width}} \qquad (8.5)$$

## FC (Initial) Example

A room that measures 24 feet by 15 feet, with a ceiling height of 9 feet 6 inches, contains eight 2-lamp, 2×4 recessed troffers. Each troffer uses reduced-wattage 4-foot lamps rated at 2700 lumens. Room surface reflectances are as follows: ceiling 70%, walls 50%, and floor 20%. Calculate the initial light level at a work plane 30 inches above the floor.

Step 1. Calculate the height of the room cavity (Hrc).

$$\text{Hrc} = 9.5 \text{ feet} - 2.5 \text{ feet} = 7 \text{ feet}$$

Step 2. Calculate the room cavity ratio (RCR), using the formula in Equation 8.6. The constant 5 is used to provide values of RCR between 1 and 10.

$$\text{RCR} = \frac{5 \times \text{Hrc} \times (\text{length} + \text{width})}{(\text{length} \times \text{width})} \qquad (8.6)$$

$$\text{RCR} = \frac{5 \times 7 \times (24 + 15)}{(24 \times 15)} = \frac{35 \times 39}{360} = 3.8$$

Step 3. Look up the CU in the CU table for the troffer (or one with similar photometrics) in the IESNA *Lighting Handbook* using the reflectances given above and the calculated RCR of 3.8. The CU for this luminaire is 0.46, found by linear interpolation of the values of CU for RCR of 3 and for RCR of 4.

Step 4. Calculate the average initial light level using Equation 8.5.

$$\text{FC (initial)} = \frac{\text{\# of fixtures} \times \text{\# of lamps/fixture} \times \text{lumens/lamp} \times \text{CU}}{\text{length} \times \text{width}}$$

$$\text{FC (initial)} = \frac{8 \times 2 \times 2700 \times 0.46}{24 \times 15} = \frac{19872}{360} = 55 \text{ FC}$$

To calculate the maintained, or design illuminance, a light loss factor (LLF; see Fact Sheet 5 and the glossary) is calculated by first multiplying together all relevant factors to obtain a LLF, sometimes called a maintenance factor. Maintained (design) illuminance is calculated using Equation 8.5.

$$\text{FC (maintained)} = \text{FC (initial)} \times \text{LLF} \qquad (8.7)$$

## FC (Maintained) Example

Ballast factor = 0.85
Lamp lumen depreciation = 0.90
Luminaire dirt depreciation = 0.92
Room surface dirt depreciation = 0.95
LLF = 0.85 × 0.90 × 0.92 × 0.95 = 0.75

FC (maintained) = 55 × 0.75 = 41 FC

To calculate the quantity of fixtures required to provide a particular illuminance (light level), Equation 8.5 is rearranged in the form of Equation 8.8.

$$\text{\# of fixtures} = \frac{\text{FC} \times \text{length} \times \text{width}}{\text{\# of lamps/fixture} \times \text{lumens/lamp} \times \text{CU}} \qquad (8.8)$$

Zonal-cavity calculations may be done with a calculator, a lighting calculator (one that is preprogrammed with the formulas in this section, such as the Halo calculator), or by using a computer program, such as Calcu-Lite (see Chapter 6).

### Spacing and Spacing Criteria

To determine how far apart luminaires may be spaced (center-to-center) to maintain uniformity, the spacing criteria values from the luminaire photometric report are used. To use the spacing criteria, multiply the net mounting height (luminaire to work plane) by the spacing criteria value. The resulting value represents the maximum (center-to-center) distance that the luminaires may be spaced to maintain uniformity.

## S.C. Example 1

A room that measures 24 feet by 15 feet, with a ceiling height of 9 feet 6 inches, contains eight 2-lamp, 2×4 recessed troffers. The spacing criteria in the lengthwise direction (obtained from the luminaire catalog) is 1.0, and in the widthwise direction it is 1.2. The work plane is at 30 inches (2.5 feet) above the floor.

Step 1. Calculate the height of the room cavity (Hrc).

$$\text{Hrc} = 9.5 \text{ feet} - 2.5 \text{ feet} = 7 \text{ feet}$$

Step 2. Calculate the spacing in the lengthwise direction.

$$\text{Sl} = 1 \times 7 = 7 \text{ feet}$$

Step 3. Calculate the spacing in the widthwise direction.

$$\text{Sw} = 1.2 \times 7 = 8.4 \text{ feet}$$

Lighting professionals who lay out luminaires use the spacing criteria in the way shown in the example; however, auditors usually use the spacing criteria differently. Working with existing lighting systems, auditors measure the center-to-center distance (S) and use the Hrc to calculate the spacing criteria. This value is then checked with the catalog value to determine if the layout exceeds the maximum spacing set by the spacing criteria.

## S.C. Example 2

An HID lighting system in a warehouse is mounted at 17 feet above the floor on 25-foot centers. If the spacing criteria of this symmetrical distribution luminaire is 1.6, are the luminaires within the maximum centers, taking the floor as the task height?

Step 1. Calculate maximum center spacing (S), based on the catalog spacing criteria value.

$$\text{S} = 1.6 \times 17 = 27 \text{ feet}$$

Step 2. Compare the existing centers with the calculated centers. If the existing (measured) center is less than the calculated center, the spacing criteria is *not* violated.

$$S \text{ calculated} = 27 \text{ feet}$$

$$S \text{ measured} = 25 \text{ feet}$$

Since S measured < S calculated, the spacing criteria is not violated.

## Lamps per Year

A standardized way of calculating the average number of lamps used in a year is useful for calculating annual lamp costs and annual labor costs. Equation 8.9 shows the calculation:

$$\text{Lamps per year} = \frac{\#\ \text{lamps/fixture} \times \text{fixtures} \times \text{burn hours/year}}{\text{average rated lamp life}} \quad (8.9)$$

Note that the calculation will be more accurate if the rated life value used is associated with the actual burn cycle of the lamp instead of the lamp catalog value of 3 hours per start.

## Economic Payback

Simple payback is a commonly used screening statistic that is simple to use, but it does not take into account the time value of money or the length of time a project will be in use. To calculate simple payback (SPB), the cost of the project is divided by the annual savings to obtain the number of years in which the project will "pay back," as shown in Equation 8.10:

$$\text{SPB} = \frac{\text{cost of project}}{\text{annual savings}} \quad (8.10)$$

Example:

$$\text{SPB} = \frac{\$9000}{\$4000} = 2.25 \text{ years } (2.25 \times 12 = 27 \text{ months})$$

## Fact Sheet 14: Lighting Maintenance

### Principles of Lighting Maintenance

1. The light output of lighting systems decreases over time.
2. Many lighting systems are overdesigned to compensate for future unnecessary light loss.
3. Improving maintenance practices can reduce light loss and can allow reductions in energy consumption or improve light levels.
4. Group maintenance practices save money.
5. Proper maintenance is the most neglected, most cost-effective way of reducing overall cost of lighting.
6. When maintenance is not performed, performance suffers gradually. The final result is a degraded lighting system performing as low as 50% of its capability.

### Effect of Maintenance on Efficiency and Effectiveness

Unlike other building energy systems, when maintenance is not performed on lighting systems, energy is reduced. The energy input to the system declines as lamps burn out and ballasts stop functioning. The system becomes less effective as light output is reduced by four recoverable light loss factors: lamp burnout factor, lamp lumen depreciation, luminaire dirt depreciation, and room surface dirt depreciation. These light loss factors can be recovered by performing maintenance. Figure 8.13 shows how neglected lighting systems lose efficiency over time.

### Group Relamping

The benefits of changing all the lamps in a group at scheduled intervals has several advantages:

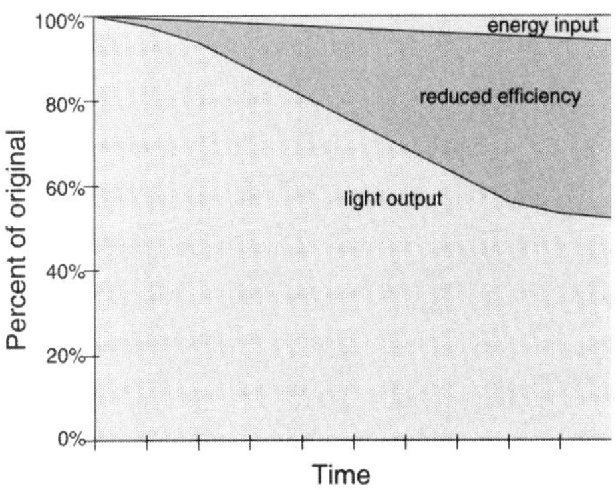


**FIGURE 8.13** Neglected lighting systems lose their efficiency over time. (*Source*: EPA Green Lights.)

1. Fewer work interruptions
2. Lower labor cost
3. Fewer lamps to stock
4. Higher maintained light level

### Action Steps

1. Schedule relamping in groups and spot relamp between intervals to reduce lumen depreciation and to save labor cost.
2. Clean fixtures at the time of relamping.
3. Write a lighting maintenance policy.
4. Design new and retrofit projects to incorporate scheduled maintenance.
5. Get help from lighting management companies.

### Troubleshooting

The best method of troubleshooting lighting is swapping known good components with unknown. We have all done this at home with incandescent bulbs. To determine if a bulb is burned out, the bulb is removed from its socket and swapped for a known good bulb (one that was seen burning in another socket). If the swapped bulb burns when screwed into the socket that held the questionable bulb, then the bulb in question must be burned out. If the good bulb does not light, the problem is in the socket or with the power. The same swapping technique can be used with fluorescent tubes to determine if the tubes or the ballasts are faulty.

Experience is the best troubleshooting tool, but there are several good troubleshooting guides available, including the IESNA *Lighting Handbook*, 8th edition (Chapter 32) and *Applied Illumination Engineering*, by Jack L. Lindsey, FIES, Fairmont Press (Chapter 17). The major lamp manufacturers also publish troubleshooting guides.

### Observing and Recording Maintenance Condition

Auditors need to identify and record the maintenance condition of lighting systems and the spaces they light. Counting the number of burned-out bulbs and tubes, the number of dark fixtures, and the dirt condition of fixtures and room surfaces is necessary to determine the state of maintenance for a facility. In most cases, maintenance is either being done or it is not, and the sharp-eyed auditor can spot the telltale signs of dirt tracks to HVAC return grills as evidence of lack of maintenance. In some cases, when energy savings must be justified, such as in performance contracting, lack of maintenance may hinder or completely obstruct the economic payback of lighting upgrades.

## Fact Sheet 15: Lamp and Ballast Disposal

Concern for environmental quality has caused lighting equipment to come under scrutiny. State and local regulations control the disposal of lamps and ballasts to keep these toxic materials out of the groundwater table.

### Lamp Disposal

All discharge lamps, fluorescent and HID, contain mercury. Mercury is a silvery-white poisonous metallic element that is liquid at room temperature. The environmental concern is for mercury released to the environment. A 4-foot, 40-watt T12 tube may contain from 27 to 48 mg of mercury, depending on the year of manufacture. Older tubes contain more mercury. Low-mercury fluorescent lamps that pass the TCLP test (Toxicity Characteristic Leachate Procedure) started to come on the market in 1996. Using these lamps avoids the requirement to treat spent lamps as hazardous material (HazMat) in most states.

**Do not throw spent lamps in the trash. Also, do not crush tubes, because mercury can become airborne and enter a worker's breathing zone.** Tubes or bulbs that are accidentally broken should be stored in sealed containers and labeled for a recycler as "Broken Fluorescent Lamps."

A TCLP test determines, in most states, whether a lamp needs to be treated as HazMat. Most fluorescent, mercury, and metal-halide lamps will fail the TCLP test. Owners of the spent bulbs and tubes are called generators and are responsible for proper disposal. Recyclers can be contracted to pick up bulbs and tubes at commercial and industrial facilities and transport them to recycling centers. Costs vary, depending on quantity. Disposal of fluorescent tubes costs about $.125 per foot.

The benefits of recycling spent lamps include the reduction of waste sent to a landfill and the promotion of materials reuse. The soda-lime glass is used in the manufacture of fiberglass or as a filler in the paving material "glass-phalt." Aluminum end caps are sent to the secondary market, like aluminum and beer cans. Mercury can be reused, after distillation, in thermometers and switches.

### Ballast Disposal

Older (core and coil) magnetic ballasts, manufactured before 1979, contain polychlorinated biphenols (PCBs). About 1 ounce of PCBs is contained as a dielectric in the ballast capacitor. Capacitors can rupture, leaching PCBs and contaminating the potting compound. The identification of PCB ballasts is negative; that is, ballasts manufactured after 1979 contain the label "NO PCBs." If a ballast has no such label and it was manufactured before 1979, there is a high probability that the ballast capacitor contains PCBs. Date codes stamped on the bottom of the ballast case can help the auditor determine when the ballast was manufactured.

PCBs are regulated at the Federal level by the TSCA (Toxic Substances Control Act) and CERCLA (Superfund Amendments). Approximately half the states regulate PCBs to varying degrees. The auditor is advised to check state regulations for more detailed information. Recycling is the best method of PCB ballast disposal. The cost is about $4 to $6 per ballast.

# Chapter 9
# Lighting Energy Conservation Measures (ECMs)

Lighting energy conservation measures (ECMs) or energy conservation opportunities (ECOs) are measures or opportunities known from actual practice to save lighting energy and reduce electric bills. They are listed in this section as general guidance for the beginning auditor, but this is not an exhaustive list of every measure known. The ECMs listed are suggested possibilities for upgrading lighting to save energy. These measures should not be applied without careful evaluation to determine whether the measure is appropriate for specific applications.

## Payback

The issue of payback invariably arises in conjunction with a discussion of ECMs. In order for an ECM to receive consideration, it must have a payback during its lifetime. Payback is a simplified way of determining how long it will take for an ECM to pay back and is calculated by dividing the cost of the ECM by the annual savings generated by the ECM. The simple payback calculation is described in detail in Chapter 8, Fact Sheet 12. While life-cycle costing (LCC) is a better measure for economic evaluation of low-risk energy investments because it considers the size and life of ECMs, simple payback is easy to understand and is widely used.

## Cost versus Persistence

There is a linear relationship between the cost of an ECM and its persistence. **Persistence** is used here in the way it is used in demand-side management (DSM) language (see Chapter 1) and means how long a measure will last or how effective it will be in reducing energy. Figure 9.1 shows how low-cost measures tend to have low persistence and more expensive measures tend to have higher persistence.

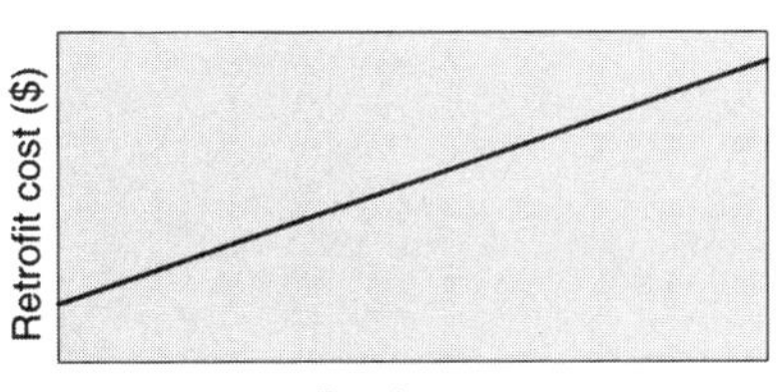


**FIGURE 9.1** Retrofit cost versus persistence.

The examples of lighting ECMs presented in this chapter are listed by cost and associated persistence. The examples range from inexpensive, low-persistence measures to higher-cost, permanent measures. Each measure is be described in detail, along with related payback information.

## Inexpensive, Low-Persistence ECMs

### Delamping Strategy

**Warning**: This strategy is *obsolete* and is usually not appropriate for most modern applications. Where applied, this strategy must not lower light levels below recommended values.

*Energy-saving strategy*: Removing lamps decreases the lighting load, reducing kWh.

*How the strategy works*: Fluorescent lamps usually operate in pairs and thus must either be removed in pairs or else some type of impedance device, such as a dummy tube or Thrift-Mate unit, must be used in place of one of the lamps to keep the other lamp operating.

*Potential savings*: Savings are proportional to the number of lamps removed. Savings are calculated by multiplying the reduced wattage by the annual burn hours and dividing by 1000. The resulting kWh value is then multiplied by the average cost per kWh (for example $.10/kWh) to obtain annual savings.

*Payback*: Less than 1 year

*Advice*: Delamping is not usually done unless there is severe overlighting, and then it is done only as a temporary measure. "Spotty" lighting usually results, lowering appearance standards. **This strategy must not lower light levels below recommended values.**

### Lamp Socket Replacements (Twist-of-the-Wrist—T.O.W.—Retrofits)
**Energy-Efficient Incandescent Bulbs:**

*Energy-saving strategy*: Energy-efficient incandescent bulbs are reduced-wattage bulbs with the same light output as standard incandescent bulbs. Energy is saved by reducing wattage.

*How the strategy works*: One class of reduced-wattage lamps uses krypton gas. Another class of reduced-wattage lamps is tungsten-halogen, which uses a halogen gas to achieve the same light output as standard incandescent bulbs while using fewer watts.

*Potential savings*: A 75-watt standard bulb can be replaced with a 52-watt tungsten-halogen bulb, saving 23 watts per socket. Operating at 3000 burn hours per year, the savings are 69 kWh per socket per year. At \$0.10/kWh, annual savings of \$6.90 per socket are realized.

*Payback*: 4 months to 1 year

*Advice*: Additional savings will be realized from maintenance savings due to the longer rated life of the tungsten-halogen bulbs. The lower-wattage replacement lamps will also reduce the cooling load (and increase the heating load), reducing air-conditioning costs (and increasing heating costs). Krypton lamps will provide less savings, but they are also less expensive to purchase. Such lamps are somewhat difficult to find due to declining manufacture.

### Energy-Efficient Incandescent Bulbs: R and PAR Bulb Replacements

*Energy-saving strategy*: Regular incandescent versions of these directional sources (spots and floods) are inefficient. A newer, more efficient version can be easily substituted as a socket replacement, saving energy by using a reduced-wattage lamp that produces the same light output and beam pattern.

*How the strategy works*: Reflector bulbs (R) are blown of soft (soda lime) glass and have a built-in parabolic reflector with a frosted coating on the face of the bulb to control the beam spread. Parabolic aluminized reflector bulbs (PAR) are manufactured of hard (borosilicate) glass and have a built-in parabolic reflector. The shape of the glass face determines the beam spread (spot or flood). Most inefficient versions are no longer manufactured due to provisions of the EPAct (see Chapter 1).

*Potential savings*: Replacing a 150-watt PAR38 standard incandescent bulb with a 75 watt PAR38 tungsten-halogen bulb of the same beam shape will provide the same light output, and will save 75 watts per socket. Operating at 4380 burn hours per year (typical for outdoor applications), the savings are 329 kWh per socket per year. At \$0.10/kWh, annual savings of \$32.90 per socket are realized.

*Payback*: Less than 1 year

*Advice*: A 75R30 incandescent bulb can be replaced with a 50PAR30LN in downlight applications, saving 25 watts per socket. Elliptical reflector (ER) and bulbous reflector (BR) bulbs can also be used in downlight applications and are exempt from EPAct. Additional savings are realized from maintenance savings due to the longer rated life of the tungsten-halogen bulbs. The lower-wattage replacement lamps will also reduce the cooling load (and increase the heating load), reducing air-conditioning costs (and increasing heating costs).

### Compact Fluorescent Lamps (CFLs)

*Energy-saving strategy*: CFLs use only one-fourth to one-third the electrical energy to produce the same lumen output as equivalent incandescent bulbs.

*How the strategy works*: An integral-ballast compact fluorescent (one-piece) has the required ballast built into the base of the bulb. The conversion from incandescent to CFL is simply a matter of removing the incandescent bulb and replacing it with a CFL of a similar shape (but note that the appropriate CFL is usually physically larger).

*Potential savings*: When a CFL is used as a socket replacement for an incandescent bulb, 66 to 75% of the original wattage is reduced, saving the same percentage of kWh and cost.

*Payback*: 1 to 4 years, depending on hours of use (see Figure 9.2)

*Advice*: CFLs have multiple sensitivities that reduce their light output. Position, thermal, and lumen depreciation sensitivities must all be accounted for when choosing a CFL on the basis of lumen equivalency. A rule of thumb is to use one wattage size larger than the raw lumen equivalent size to account for these effects, unless amalgam lamps are used.

Physical size may not allow the use of a CFL in all cases. CFLs should not be used for spotlighting or highlighting since they are linear, not point sources. Tungsten-halogen sources work best for these applications.

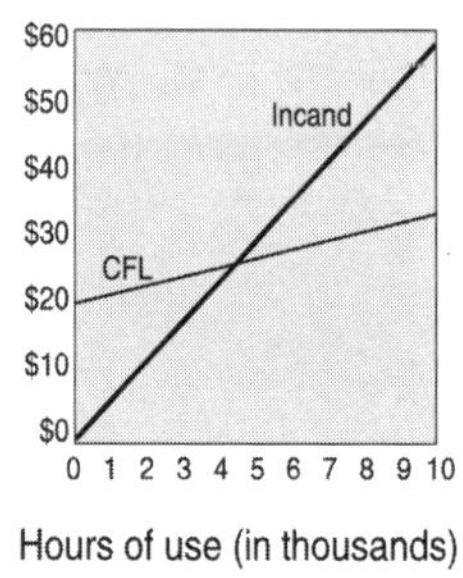


**FIGURE 9.2** Total cost for 10,000 hours of use for incandescent and CFL bulbs.

### T10 Q&D ("Quick and Dirty") Fixture Upgrade

*Energy-saving strategy*: Using 4-foot T10 lamps and delamping reduces the power input to a fixture, with only a slight lowering of the light level, resulting in a "quick and dirty" (Q&D) retrofit.

*How the strategy works*: T10 lamps are high-lumen lamps (3500-3700 lumens) that may be operated on T12 ballasts and are used in place of two of the four 40-watt (or 34-watt) 4-foot fluorescent tubes in a 4-tube fixture, using one of the existing magnetic ballasts. The other ballast may be removed.

*Potential savings*: See Table 9.1 for an example of savings that can be realized when delamping a 4-tube fixture and relamping with two T10 lamps.

**TABLE 9.1** Delamping 4-lamp T12 Fixtures and Replacing with 2 T10 Lamps (with 1 Std-Mag Ballast)*

| | 4 T12 Lamps with 2 Std. Magnetic Ballasts | |
|---|---|---|
| | 40-W Lamps | 34-W Lamps |
| Savings | 95 | 63 |
| Hours/year | 3000 | 3000 |
| kWh/yr savings | 285 | 189 |
| Avg. $/kWh | $0.10 | $0.10 |
| Savings/yr | $28.50 | $18.90 |

**Fixture wattage for T10s is 95.*

*Payback*: Less than 2 years, depending on hours of use

*Advice*: The T10 lamp actually consumes 42 watts on a magnetic ballast, but this has been considered in the calculations above. Remove and recycle the unused magnetic ballast. Either the inboard or outboard pair of lamps may be removed, but be consistent so all fixtures appear the same. Many lighting professionals prefer to leave the outboard pair. Cleaning of the white paint reflector and the lens is recommended. Beware of purchasing agents who argue that the T10 tube is too expensive (approximately $7 each, in quantity). With an energy payback of less than 2 years, this strategy is cost-effective. The T10 tube also has a longer average rated life (24,000 hours), so maintenance costs are reduced.

### EXIT Light Retrofit Lamps

Energy-saving strategy: Special retrofit bulbs designed to replace standard EXIT bulbs are used. The retrofit bulbs are lower wattage, reducing kWh.

*How the strategy works*: There are several types of replacement bulbs. Some are incandescent and consist of a series of miniature lamps operated undervoltage for long life (30,000 to 130,000 hours). A variety of bases accommodates many different exit light lamp sockets. Some brand names are Eternalamp and Flex-Lite. Some EXIT lights, such as the Merlin by Lithonia Lighting, are tubes with arrays of LEDs (light-emitting diodes),

*Potential savings*: Two 9-watt Eternalamps can produce energy savings of $19/unit/year. Two 1-watt Merlin bulbs can produce energy savings of $33/unit/year. Estimates are made using a cost of $.10/kWh, and savings are based on comparison with two 20-watt incandescent bulbs. There would also be additional maintenance labor savings.

*Payback*: 1 to 2 years

*Advice*: Evaluate the condition of the old EXIT light before purchasing retrofit lamps. New units may be a better option. The National Fire Protection Association (NFPA) Life Safety Code and the National Electric Code (NEC) must be met. NEC requires "2 elements," so more than one lamp is required. UL924 applies to new EXIT lights and retrofit kits and is a worthy goal, especially with regard to the luminance uniformity requirement. UL924 requires a minimum face luminance, measured in numerous locations to ensure that the exit sign can be seen.

Some retrofit kits require a change of diffuser to meet the face luminance uniformity.

If LED "tubes" are selected, beware of designs that employ circuits that overdrive the LEDs and/or do not protect the LEDs against voltage transients. LED end-of-life and lumen depreciation are accelerated when overdriven.

## Medium-Cost, More Permanent ECMs

### T8/ELB Fixture Upgrade

*Energy-saving strategy*: The T8 lamp with electronic ballast (ELB) is a high-efficiency combination.

*How the strategy works*: This lamp replacement for the older T12 lamp it replaces is more efficient (produces more lumens per watt) than the T12 lamp. A ballast change is required, and the most efficient lamp/ballast combination is obtained with a high-frequency electronic ballast.

*Potential savings*: Changing from a 4-lamp, T12 fixture to four T8 lamps with one 4-lamp instant-start electronic ballast will reduce the fixture wattage from 190 to 105 watts, saving 85 watts per fixture. Operated at 3000 hours per year, the savings in energy is 255 kWh per fixture per year. At $.10/kWh the annual savings is $25.50 per fixture.

*Payback*: 2 to 3 years, depending on hours of use and installation costs

*Advice*: The lamp is smaller in diameter compared with the T12 it replaces, with the T8 measuring 1 inch in diameter, compared to the 1.5-inch diameter of the T12. Both lamps have the same pin-spacing, so the T8 lamp will mechanically fit into a socket formerly occupied by a T12 lamp. A ballast change is required to properly operate the T8 lamp. Multilamp electronic (high-frequency) ballasts are the most efficient combination (4 lamps max). If lamps are loose in the sockets, do not reuse the sockets, but replace with new sockets or consider new fixtures instead of a retrofit of existing fixtures.

## EXIT Light LED Kit

*Energy-saving strategy*: Solid-state light sources reduce constant-burn lighting load, resulting in high kWh savings.

*How the strategy works*: LEDs are solid-state light sources that consume 96% less energy than incandescent bulbs and 85% less energy than compact fluorescents (CFLs). EXIT lights are constant-burn applications (8760 hours/year), so the potential for savings is high. The rated life of LEDs is 80 years, so EXIT light maintenance costs are eliminated. Each kit consists of a new diffuser and two strips of LEDs that mount inside the EXIT light.

*Potential savings*: If existing exit lights use two 20-watt incandescent lamps, savings of 95% can be realized. At $.10/ kWh, $33 per EXIT light per year is saved in electricity costs alone. Maintenance savings would add to these savings.

*Payback*: 1 to 2 years

*Advice*: Ensure that the kit meets or exceeds building fire codes (NFPA), NEC, and UL listing that apply to EXIT sign lighting. To achieve long life, beware of designs that employ circuits that overdrive the LEDs and/or do not protect the LEDs against voltage transients.

## *In Situ* Luminaire Upgrades

*Energy-saving strategy*: Replacing one or more luminaire components currently in place in order to increase efficiency and decrease the lighting load, in watts.

*How the strategy works*: Changing components such as lamps and ballasts, adding reflectors, and replacing a lens with a louver are all examples of luminaire efficiency improvements. Replacing these components in place (in situ) is called a retrofit and is done with field labor.

*Potential savings*: 25 to 50%

*Payback*: 2 to 3 years, depending on hours of use, extent of retrofit, and electric rate

*Advice*: T8 and CFL lamps: Watch the effects of concentrated brightness and CRI (color rendering index) on visual perception.

*Electronic ballasts*: Watch ballast factor, THD (total harmonic distortion), and in-rush current from low THD ballasts.

*Sockets*: If sockets are not in good shape, replacement may be too costly in situ and new luminaires may be less expensive. See "New Luminaires (Relighting)," which follows this section.

*Reflectors*: Watch effects on spacing criteria such as nonuniform lighting resulting in severe wall shadows and dark area between fixtures.

*Economics*: Compare retrofit with the relighting option. Extensive retrofits may be more costly than relighting. See "New Luminaires (Relighting)," which follows this section.

## New Luminaires (Relighting)

*Energy-saving strategy*: Replacing a lighting system with new, energy-efficient luminaires to decrease the lighting load.

*How the strategy works*: Relighting provides the opportunity to change the variables of the system, including the type and number of lamps per luminaires, the type and number of luminaires, and the layout. For example, an existing system of 4-lamp, T12 2×4 luminaires may oftentimes be replaced with 3-lamp, T8 2×4 luminaires. A single, high-frequency electronic ballast would probably be specified in each luminaire to improve both efficiency and quality (by decreasing flicker). An alternative would be 2-lamp tandem-wired electronic ballasts, which would allow inboard/outboard switching to provide multiple light levels.

Relighting also affords the opportunity to upgrade lighting quality by providing better shielding: for example, replacing lensed troffers with deep-cell parabolic luminaires.

*Potential savings*: 25 to 60%

*Payback*: 2 to 4 years, depending on hours of use, electric rate, and performance of old system

*Advice*: Alternatives that work especially well in computer environments include indirect or direct-indirect luminaires. Select luminaires with visual comfort probability (VCP) greater than 80.Consider adding controls to turn lights off when not needed or to dim lights when tasks change. Choose room surfaces that are lighter (and have a matte finish), especially in the case of ceilings and partitions.

## Task/Ambient Lighting System

*Energy-saving strategy*: Applied in office and industrial lighting and consisting of general uniform, low-level illumination from an overhead ambient system, supplemented with a higher level of task illumination at workstations.

*How the strategy works*: The lower ambient light level can be achieved with less power, resulting in a lower total power density than with comparable direct systems alone. The supplementary task luminaires may be switched off when not required.

*Potential savings*: 25 to 40%

*Payback*: 2 to 4 years, depending on hours of use, electric rate, and performance of old system

*Advice*: Ensure that walls are adequately lighted to prevent the cavelike appearance that accompanies dark walls. Select efficient, low-glare task lights, and watch for veiling reflections on work surfaces.

## Switch/Timer

*Energy-saving strategy*: Timed control turns off lights.

*How the strategy works*: Preset electronic timers turn off lights. User sets time from 10 minutes to 12 hours in the switch-plate version or from 1 minute to 18 hours in the wall switch version. Turning on the lights starts the timer, and if the lights are not turned off during the timing period, the lights will be turned off att he end of it. The timer is reset by turning the switch off. Options include a three-way switch version and a model that "flicks" the lights as a warning to notify the user to reset the timer if more time is needed.

*Potential savings*: No savings if occupants turn off lights religiously. However, in situations where lights are often left on inadvertently, savings of 10 to 50% may be realized.

*Payback*: 1 to 3 years

*Advice*: Good alternative to occupancy sensors in irregularlyoccupied spaces. For safety, the models with a warning "flick" are recommended.

## Astronomical Clock Timer

*Energy-saving strategy*: Turning outside lights on at dusk and off at dawn.

*How the strategy works*: A special timer tracks sunrise and sunset electronically. An additional channel allows timed control of lights at fixed times. Lights turned on automatically at sunset can be turned off at a time between dusk and dawn.

*Potential savings*: Savings realized with the clock timer beyond those obtainable with a photocell come (1) from having a higher reliability than a photocell, since photocells fail ON and are not always replaced when they fail, keeping lights on continuously, and (2) from the ability to program a time off before dawn. With a timer, 20 to 50% savings are common. Savings are usually off-peak.

*Payback*: 2 to 5 years

*Advice*: Installation is easiest when the unit is installed in a junction box near the circuit-breaker box that feeds the lighting load. Programming directions are easy to follow and a battery backup or super-cap holds the program if power is interrupted.

## Occupancy Sensors

*Energy-saving strategy*: Turning off lights in unoccupied spaces when there is no occupancy.

*How the strategy works*: A sensor turns lights on when a person enters a controlled area and off after the area is vacated. Two main technologies are used: passive infrared (PIR) and ultrasonic. PIR reacts only to energy sources, such as the human body, and must be able to "see" an area to control it. PIR sensors sense occupancy by detecting the difference between the heat source and the background space. Ultrasonic sensors emit high-frequency sound waves (above the range of human hearing) and measure the time the signals return to the sensor. Movement by a person causes the sound waves to return at a different rate, resulting in a shift that is detected by the sensor.

*Potential savings*: No savings if occupants turn off lights religiously; however, research has shown that single-person offices are only occupied for 4 hours a day, but the lights are left on as much as 24 hours a day, 7 days a week. Savings of 30% to 50% have been measured after installing sensors.

*Payback*: 2 to 3 years, depending on lighting load and hours saved

*Advice*: Survey building to locate spaces where lights are left on when there is "no one home." Offices, restrooms, conference rooms, classrooms, and warehouses are excellent candidates. Choose correct technology (PIR or ultrasonic) for the application. Remember that PIR cannot "see" behind toilet stalls or partitions. Wall box units are not universally applied. Proper placement of sensors is the key to successful application. Ultrasonic sensors need to be located away from heating and cooling diffusers.

## Lighting Maintenance Considerations—*No* Energy Payback

A common energy conservation measure often found on ECM lists for heating, ventilation, and air-conditioning (HVAC) applications is "equipment maintenance." When maintenance is performed on HVAC systems, building controls, and motors, energy is saved, usually due to lower horsepower requirements, which lower the kWh consumed. However, unlike for HVAC, controls, and motors, **for lighting systems energy costs are not saved when systems are maintained but costs are actually saved when systems are neglected.**

Neglected lighting systems lose efficiency over time (see Chapter 8, Fact Sheet 14, and Figure 8.13). As lamps burn out and ballasts need to be replaced, the energy input of the system declines, reducing kWh. Light levels also decline over time due to lamp burnouts, lamp lumen depreciation, and dirt depreciation. These energy and light level reductions occur at the expense of employee productivity and morale—intangible paybacks for some companies.

# Chapter 10
# Checklist Forms

The use of preprinted survey and audit forms is described in Chapter 5. Preprinted forms are advantageous because they provide (1) a preset scope of data collection, (2) consistent data collection with low incidence of missing data, and (3) ease of reviewing and reconstructing data. When all surveyors or auditors use the same forms, data collection problems are minimized.

The team notebooks containing the preprinted audit forms are filled out as the teams move through the building, using one form for each space. Keeping these forms in the same order as the teams go through the building has proved valuable to fill gaps in information discovered when doing the computer entries on-site. Revisiting some areas is easier while still on-site than revisiting the site later! Also, questions that arise during the audit can be directed to the client's representative, and the equipment or condition can be viewed and discussed firsthand.

Several types of survey and audit forms are provided in this chapter. They are provided for field surveyors and auditors who are encouraged to copy or modify them to fit their needs. These forms have been field-tested and have proved to provide efficient, consistent data collection.

## Walk-Thru Survey Forms

### Walk-Thru Lighting Survey Checklists

Checklist 10.1 is a lighting survey checklist used for a walk-thru survey to determine retrofit potential. Light levels are assessed visually and are the impressions of the surveyor, judged to be either "OK" or "Not OK." The list of one-for-one retrofit options includes energy conservation measures (ECMs) described in Chapter 9, such as Twist-of-the-Wrist (T.O.W.) lamp changes, lamp and ballast upgrades, increasing luminaire efficiency, and complete system change-out (relighting). Controls are treated separately, and recommendations are made for adding controls where none currently exist or replacing one type of control for another (for example, replacing a wall switch with an occupancy sensor). Several lines are provided for recording comments, such as the state of maintenance and depreciation factors. One form is used for each office or area of the building.

A step-by-step guide for the beginning surveyor to calculate savings and payback is provided in Figure 10.2 and may be copied for inclusion in the surveyor's notebook.

### Walk-Thru Lighting Survey Form

Checklist 10.3 provides a more detailed lighting survey form that can be enlarged for ease of use. This form is used when fixtures are counted and described. The power consumption of each fixture is estimated, and recommended changes are made to upgrade each fixture type. The "Power Reduction" column is used to calculate the difference between the original fixture power and the power after the retrofit. The estimated hours of use are used to multiply values (in kW) in the "Power Reduction" column to obtain the values in the "Annual kWh Savings" column, and then the "Estimated Annual Savings/Fix" is calculated by multiplying values in the "Annual kWh Savings" column by the average cost per kWh and dividing by values in the "Qty" column. The form provides a column for recording estimated material costs (per fixture) that can be used to calculate simple payback (also on the form). The last two columns are provided for recording comments and priority-ranking the retrofit measure.

## Lighting Audit Forms

### Office Lighting Checklist

The Office Lighting Checklist shown in Figure 10.4 is a form used for intermediate and comprehensive audits of office lighting. One form is used for each office or open office area. The form is copied two-sided, with page **00** showing the front side and page **00** showing the back side.

The first section is for recording general information about the office: its dimensions, surface finishes, cleanliness, and illuminance readings. The data collected includes the following:

1. Building identification and/or office ID that contains the name of the building and/or office; if a person's name is on an office, this information can be entered
2. Date of the audit
3. Names or initials of auditors

## LIGHTING SURVEY CHECKLIST

Area ______________________________

| | | | OK | NOT OK |
|---|---|---|---|---|
| 1. | Light level | | ____ | ____ |

2. Retrofit: T.O.W. Lamp change ______________________________

3. Retrofit: Upgrade lamps & ballast ______________________________

4. Retrofit: Increase luminaire efficiency ______________________________

5. Retrofit: System change ______________________________

| | | | Add | Change |
|---|---|---|---|---|
| 6. | Controls: | ______________________________ | ____ | ____ |

Comments:

______________________________

______________________________

______________________________

______________________________

______________________________

Checklist 10.1

## Lighting Walk-Thru Survey
## Calculating Retrofit Savings & Payback

A. Define existing conditions - Perform a Lighting Audit
1. Fixture watts
2. Quantity of fixtures
3. Calculate total watts (kW)
4. Burn hours (hrs)
5. Calculate annual kWh

B. Define proposed retrofit conditions
1. Fixture watts after retrofit
2. New burn hours
3. Calculate annual kWh

C. Define economics - Calculate payback
1. Determine cost to retrofit
2. Calculate Annual savings in kWh and kW

   Savings (kWh) = Old kWh - New kWh

   Savings (kW) = Old kW - New kW
3. Incremental cost of electricity per kWh
4. Incremental cost of electricity per kW demand
5. SPB = $\dfrac{\text{Cost of Retrofit}}{\text{Annual Savings}}$

(Checklist 10.2)

Lighting Survey

Site:
Address:

Date performed:

Survey conducted by:

| Qty | Existing Light Fixture Description | Est Watts | Recommended Change | Watts | Power Reduction | Est hrs per year | Annual kWh Savings | Estimated Annual Savings/Fix | Estimated Material Cost | Simple Payback (yrs) | Comments | Priority |
|---|---|---|---|---|---|---|---|---|---|---|---|---|
| | | | | | | | | | | | | |
| | | | | | | | | | | | | |
| | | | | | | | | | | | | |
| | | | | | | | | | | | | |
| | | | | | | | | | | | | |
| | | | | | | | | | | | | |
| | | | | | | | | | | | | |
| | | | | | | | | | | | | |
| | | | | | | | | | | | | |
| | | | | | | | | | | | | |
| | | | | | | | | | | | | |
| | | | | | | | | | | | | |
| | | | | | | | | | | | | |
| | | | | | | | | | | | | |
| | | | | | | | | | | | | |
| | | | | | | | | | | | | |
| | | | | | | | | | | | | |
| | | | | | | | | | | | | |
| | | | | | | | | | | | | |

**(Checklist 10.3)**

# OFFICE LIGHTING CHECKLIST

**Building ID:** ______________________ **Office ID:** ______________________
**Date:** __________ **Auditors:** __________________

**Length =** ________ **ft** **Width =** _________ **ft** **Ceiling Ht =** _________ **ft**

**FINISHES: Ceiling:** L M D **Walls:** L M D **Floors:** L M D

**AREA CLEANLINESS:** VC Clean Avg Dirty VD

**FC** (high): _______ **FC** (low): ________

---

**LAMP 1** **LAMP WATTS:** ______________ **LAMPS/FIXTURE:** ____________
**FLUOR:** CFL U Straight **LENGTH:** 8' 6' 5' 4' 3' 2'
**DIA:** T5 T8 T10 T12 T17 **Color:** CW WW Day LW RE
**INCAND:** A R PAR Other: ______________
**HID:** MH Mercury **LAMP CONDITION:** New Old Dark Ends
**BASE:** Mog Med **BULB:** Clear Coated
**DEPRECIATION:** Max Min Mixed **BURNOUTS:** __________

- - - - - - - - - - - - - - - - - - - - - - - - - - - - - - - - - - - - - -

**FIXTURE 1** **VOLTAGE:** 277 208 120 **INPUT WATTS (or Cat #)**: __________
**SHAPE:** Round Square Rect **HOUSING:** Open Enclosed
**SIZE:** 1X4 2X4 2X2 Other **CONDITION:** VD Dirty Avg Clean VC
**QTY:** ______ **OPT CTRL:** Open Lens Louver WA Rusty Peeling
**LENGTHWISE SPACING (Center-to-center):** ________ ft Cont Row
**Burnouts:** ___**WIDTHWISE SPACING (Center-to-center):** __________ft Cont Row
**SWITCHING 1** Bkr PB/Contactor Sw Tmr **BURN HOURS:** Hrs/Wk: ___Wks/Yr:____

---

**LAMP 2** **LAMP WATTS:** ______________ **LAMPS/FIXTURE:** ____________
**FLUOR:** CFL U Straight **LENGTH:** 8' 6' 5' 4' 3' 2'
**DIA:** T5 T8 T10 T12 T17 **Color:** CW WW Day LW RE
**INCAND:** A R PAR Other: ______________
**HID:** MH Mercury **LAMP CONDITION:** New Old Dark Ends
**BASE:** Mog Med **BULB:** Clear Coated
**DEPRECIATION:** Max Min Mixed **BURNOUTS:** __________

- - - - - - - - - - - - - - - - - - - - - - - - - - - - - - - - - - - - - -

**FIXTURE 2** **VOLTAGE:** 277 208 120 **INPUT WATTS (or Cat #)**: _________
**SHAPE:** Round Square Rect **HOUSING:** Open Enclosed
**SIZE:** 1X4 2X4 2X2 Other **CONDITION:** VD Dirty Avg Clean VC
**QTY:** ______ **OPT CTRL:** Open Lens Louver WA Rusty Peeling
**LENGTHWISE SPACING (Center-to-center):** ________ ft Cont Row
**Burnouts:** ___**WIDTHWISE SPACING (Center-to-center):** __________ft Cont Row
**SWITCHING 2** Bkr PB/Contactor Sw Tmr **BURN HOURS:** Hrs/Wk: ___Wks/Yr:____

---

**Comments:**

**(Checklist 10.4)** Front.

4. Room dimensions: length, width, ceiling height
5. Room surface finishes: ceilings, walls, floors (identifying them as light-colored, medium-colored, or dark-colored—L, M, or D)
6. Area cleanliness: very clean (VC), clean, average, dirty, very dirty (VD)
7. Illuminance: high footcandle (FC) measurement or low FC measurement

The next two sections go together—Lamp 1 and Fixture 1. The lamp (bulb) section describes the lamp watts; number of lamps per fixture; the type of lamp; length and diameter; phosphor color (CW = cool white, WW = warm white, Day = daylight, LW = light white, RE = rare earth); base (Mog = mogul base, Med = medium base); depreciation; and percent burnouts. The fixture section describes the voltage; input watts (or ballast catalog number); shape; nominal size (1×4, 2×4, 2×2, etc.); housing; quantity of fixtures in this office; fixture optical control (open fixture, lens, louver, or wraparound [WA]); physical condition (degree of cleanliness, and whether fixture is flaking or peeling); center-to-center spacing in the lengthwise direction and the widthwise direction (or continuous rows); the type of switching (circuit breaker, pushbutton switch and contactor, wall switch, or timer—Bkr, PB/Contactor, Sw, and Tmr), and the burn hours (hours per week and weeks per year). Lamp 2 and Fixture 2 sections are provided for cases in which there is more than one lamp and/or fixture type in the same office.

The back side of the form contains a Task Lighting Checklist that is used to record information about any task lighting used in the office. Information recorded here includes the tasks performed, the mounting height of the task lighting, high and low FC measurements, and lamp and fixture information (in a format similar to that on the reverse side).

The grid of 1/4-inch squares at the bottom of the back side is provided as a sketch area for drawing any unusual ceiling layouts or task lighting arrangement. The recording auditor labels the sketch with any additional information that may be useful, such as dimensions.

### Industrial Lighting Checklist

The Industrial Lighting Checklist shown in Checklist 10.5 is a form used for intermediate and comprehensive audits of industrial lighting. One form is used for each area of a manufacturing facility. The form is copied two-sided, with page **00** on the front side and page **00** on the back side.

The first section includes the date of the audit, the auditors names or initials; the building identifications; the ceiling height; the truss height and the internal boundaries (column-to-column) of the area included for this form; manufacturing activities; ceiling, wall, and floor finishes; and the area cleanliness. Refer to the previous discussion of the Office Lighting Checklist (page **00**) for an explanation of the abbreviations used.

The next section is used to record the fixture and lamp data, mounting information (SM = surface-mount; Pend = pendant-mount), lamp and fixture depreciation, fixture condition, quantity of fixtures in this area, percent of burnouts, switching equipment, the burn hours per week, and weeks per year. This section is repeated so two fixture and/or lamp types can be recorded on one sheet for the same area. A section for comments is included at the bottom of the front page.

The back side of the form is used for recording task lighting found in the same area as the general (ambient) lighting systems.

## Fixture Schedule Forms

Checklist 10.6 is used when the audit starts with a walk-thru to identify all the different fixture types, before the intermediate or comprehensive audit is begun. A fixture and lamp combination constitutes a fixture type. Two fixture categories are used: interior and exterior. The interior fixtures are further classified by the lamp type.

The list starts with II-1, with II being the abbreviation for Interior Incandescent fixtures. The lines after each fixture type are used to describe that fixture type. Interior fluorescent fixtures are designated IF, IH refers to HID fixtures, and E designates EXIT lights. In the exterior list, EI refers to exterior incandescent fixtures, while EH means exterior HID fixtures.

After the walk-thru survey is finished, copies of the fixture schedules are placed in each audit team's notebook. The basic audit checklist forms are simplified, since the repeated lamp and fixture information does not need to appear on each sheet. Thus, only fixture identification is used, as shown in Checklists 10.7 and 10.8.

# TASK LIGHTING CHECK LIST

**TASKS:** ____________________
____________________
____________________

**MOUNTING Ht:** ____________
**FC (high):** __________ **FC(low):** __________

---

**LAMP** **LAMP WATTS:** ______________ **LAMPS/FIXTURE:** ____________
**FLUOR:** CFL U Straight **LENGTH:** 8' 6' 5' 4' 3' 2'
**DIA:** T5 T8 T10 T12 T17 **Color:** CW WW Day LW RE
**INCAND:** A R PAR Other: ______________
**HID:** MH Mercury
**BASE:** Mog Med **BULB:** Clear Coated
**DEPRECIATION:** Max Min Mixed **BURNOUTS:** __________

- - - - - - - - - - - - - - - - - - - - - - - - - - - - - - - - - - - -

**FIXTURE** **VOLTAGE:** 208 120 **INPUT WATTS (or Cat #):**___________
**SHAPE:** Round Square Rect **HOUSING:** Open Enclosed
**QTY:** _____ **OPT CTRL:** Open Lens Louver WA**CONDITION:** VD Dirty Avg Clean VC
**SIZE:** 1x4 2x4 2x2 **WIRING:** Perm Port
**LENGTHWISE SPACING (Center-to-center):** _________ ft Cont Row
**Burnouts:** __**WIDTHWISE SPACING (Center-to-center):** _________ft Cont Row
**SWITCHING:** Local Sw Remote Sw **BURN HOURS:** Hrs/Wk: ___ Wks/Yr:___

**(Checklist 10.4)** Back.

# INDUSTRIAL LIGHTING SURVEY

Date: ___________ Auditors: _______________________

Building ID: ____________________________________ Ceiling Ht = _____________ ft

Internal Boundaries: Column: _______ to Column: _______ Truss Ht = _____________ ft

ACTIVITIES: Process Assembly Testing Storage Task Other:____________

FINISHES: Ceiling: L M D Walls: L M D Floors: L M D

AREA CLEANLINESS: VC Clean Dirty VD FC (high): ______ FC (low): _______

---

FIXTURE TYPE:__________ LAMP WATTS: _____ Fluorescent LENGTH: ___ft

SIZE: __________________ COLOR: CW WW Day LW RE

MOUNTING: Recessed SM Pend Incand Mercury MH HPS

DEPRECIATION: Max Min Mixed CONDITION: VD Dirty Avg Clean VC Rusty Peeling Lens Degraded

QTY: ______ BURNOUTS: _________ SWITCHING: Bkr PB/Contactor Sw Tmr

BURN HOURS: Hrs/Wk: ____ Wks/Yr: ___

- - - - - - - - - - - - - - - - - - - - - - - - - - - - - - - - - - - - - - -

FIXTURE TYPE:__________ LAMP WATTS: _____ Fluorescent LENGTH: ___ft

SIZE: ___________________ COLOR: CW WW Day LW RE

MOUNTING: Recessed SM Pend Incand Mercury MH HPS

DEPRECIATION: Max Min Mixed CONDITION: VD Dirty Avg Clean VC Rusty Peeling Lens Degraded

QTY: ______ BURNOUTS: _________ SWITCHING: Bkr PB/Contactor Sw Tmr

BURN HOURS: Hrs/Wk: ____ Wks/Yr: ___

- - - - - - - - - - - - - - - - - - - - - - - - - - - - - - - - - - - - - - -

Comments:

## TASK LIGHTING CHECK LIST

TASKS: ______________________ AREA BOUNDARIES: ______________

MOUNTING Ht: __________ FC (high): ______ FC(low): _______

FIXTURE TYPE: _________ LAMP LENGTH: ____ ft LAMP WATTS: ______

MOUNTING: Recessed SM Pend COLOR: CW WW Day LW RE

DEPRECIATION: Max Min Mixed CONDITION: VD Dirty Avg Clean VC
Rusty Peeling Lens Degraded

QTY: _____ BURNOUTS: _______ SWITCHING: Bkr PB/Contactor Sw Tmr

BURN HOURS: Hrs/Wk: ____ Wks/Yr: ___

- - - - - - - - - - - - - - - - - - - - - - - - - - - - - - - -

Comments:

- - - - - - - - - - - - - - - - - - - - - - - - - - - - - - - -

TASKS: ______________________ AREA BOUNDARIES: ______________

MOUNTING Ht: __________ FC (high): ______ FC(low): _______

FIXTURE TYPE: _________ LAMP LENGTH: ____ ft LAMP WATTS: ______

MOUNTING: Recessed SM Pend COLOR: CW WW Day LW RE

DEPRECIATION: Max Min Mixed CONDITION: VD Dirty Avg Clean VC
Rusty Peeling Lens Degraded

QTY: _____ BURNOUTS: _______ SWITCHING: Bkr PB/Contactor Sw Tmr

BURN HOURS: Hrs/Wk: ____ Wks/Yr: ___

- - - - - - - - - - - - - - - - - - - - - - - - - - - - - - - -

Comments:

**Checklist 10.5** Back.

Checklist 10.6

# ABC CORPORATION
## FIXTURE SCHEDULE

| | | | |
|---|---|---|---|
| INTERIOR: | Incandescent | II-1 | |
| | Incandescent | II-2 | |
| | Incandescent | II-3 | |
| | Incandescent | II-4 | |
| | Fluorescent | IF-1 | |
| | Fluorescent | IF-2 | |
| | Fluorescent | IF-3 | |
| | Fluorescent | IF-4 | |
| | Fluorescent | IF-5 | |
| | HID | IH-1 | |
| | HID | IH-2 | |
| | HID | IH-3 | |
| | HID | IH-4 | |
| | EXIT | E-1 | |
| | EXIT | E-2 | |
| EXTERIOR: | Incandescent | EI-1 | |
| | Incandescent | EI-2 | |
| | HID | EH-1 | |
| | HID | EH-2 | |
| | HID | EH-3 | |
| | HID | EH-4 | |

# OFFICE LIGHTING CHECKLIST

**Building ID:** ______________________ **Office ID:** ______________________
**Date:** __________ **Auditors:** ______________________

**Length =** ________ **ft** **Width =** __________ **ft** **Ceiling Ht =** __________ **ft**

**FINISHES: Ceiling:** L M D **Walls:** L M D **Floors:** L M D

**AREA CLEANLINESS:** VC Clean Avg Dirty VD

**FC** (high): _______ **FC** (low): ________

---

**FIXTURE ID:** ______________

**LAMP DEPRECIATION:** Max Min Mixed **BURNOUTS:** __________

- - - - - - - - - - - - - - - - - - - - - - - - - - - - - - - - - - - - - - - -

**VOLTAGE:** 277 208 120

**CONDITION:** VD Dirty Avg Clean VC

**QTY:** ______ **LENGTHWISE SPACING (Center-to-center):** ________ ft Cont Row

**Burnouts:** ___ **WIDTHWISE SPACING (Center-to-center):** __________ft Cont Row

**SWITCHING 1** Bkr PB/Contactor Sw Tmr **BURN HOURS:** Hrs/Wk: ___Wks/Yr:____

---

**FIXTURE ID:** ______________

**LAMP DEPRECIATION:** Max Min Mixed **BURNOUTS:** __________

- - - - - - - - - - - - - - - - - - - - - - - - - - - - - - - - - - - - - - - -

**VOLTAGE:** 277 208 120

**CONDITION:** VD Dirty Avg Clean VC

**QTY:** ______ **LENGTHWISE SPACING (Center-to-center):** ________ ft Cont Row

**Burnouts:** ___ **WIDTHWISE SPACING (Center-to-center):** __________ft Cont Row

**SWITCHING 1** Bkr PB/Contactor Sw Tmr **BURN HOURS:** Hrs/Wk: ___Wks/Yr:____

---

**Comments:**

**Checklist 10.7** Front.

# TASK LIGHTING CHECK LIST

**TASKS:** ____________________
____________________
____________________

**MOUNTING Ht:** ____________
**FC (high):** __________ **FC(low):** __________

---

**LAMP** **LAMP WATTS:** ____________ **LAMPS/FIXTURE:** ____________
**FLUOR:** CFL U Straight **LENGTH:** 8' 6' 5' 4' 3' 2'
**DIA:** T5 T8 T10 T12 T17 **Color:** CW WW Day LW RE
**INCAND:** A R PAR Other: ____________
**HID:** MH Mercury
**BASE:** Mog Med **BULB:** Clear Coated
**DEPRECIATION:** Max Min Mixed **BURNOUTS:** __________

- - - - - - - - - - - - - - - - - - - - - - - - - - - - - - - - - - - - - - - -

**FIXTURE** **VOLTAGE:** 208 120 **INPUT WATTS (or Cat #):** __________
**SHAPE:** Round Square Rect **HOUSING:** Open Enclosed
**QTY:** _____ **OPT CTRL:** Open Lens Louver WA **CONDITION:** VD Dirty Avg Clean VC
**SIZE:** 1x4 2x4 2x2 **WIRING:** Perm Port
**LENGTHWISE SPACING (Center-to-center):** _________ ft Cont Row
**Burnouts:** __ **WIDTHWISE SPACING (Center-to-center):** _________ ft Cont Row
**SWITCHING:** Local Sw Remote Sw **BURN HOURS:** Hrs/Wk: ___ Wks/Yr: ___

**Checklist 10.7** Back.

# INDUSTRIAL LIGHTING SURVEY

**Date:** ___________ **Auditors:** ______________________

**Building ID:** ______________________________________ **Ceiling Ht =** _____________ **ft**

**Internal Boundaries: Column:** _______ **to Column:** _______ **Truss Ht =** _____________ **ft**

**ACTIVITIES:** Process Assembly Testing Storage Task Other:____________

**FINISHES: Ceiling:** L M D **Walls:** L M D **Floors:** L M D

**AREA CLEANLINESS:** VC Clean Dirty VD FC (high): _____ FC (low): _______

---

**FIXTURE ID:**___________

**DEPRECIATION:** Max Min Mixed **CONDITION:** VD Dirty Avg Clean VC
Rusty Peeling Lens Degraded

**QTY:** ______ **BURNOUTS:** _________ **SWITCHING:** Bkr PB/Contactor Sw Tmr

**BURN HOURS: Hrs/Wk:** ____ **Wks/Yr:** ___

- - - - - - - - - - - - - - - - - - - - - - - - - - - - - - - -

**FIXTURE ID:**___________

**DEPRECIATION:** Max Min Mixed **CONDITION:** VD Dirty Avg Clean VC
Rusty Peeling Lens Degraded

**QTY:** ______ **BURNOUTS:** _________ **SWITCHING:** Bkr PB/Contactor Sw Tmr

**BURN HOURS: Hrs/Wk:** ____ **Wks/Yr:** ___

- - - - - - - - - - - - - - - - - - - - - - - - - - - - - - - -

**Comments:**

**Checklist 10.8** Front.

## TASK LIGHTING CHECK LIST

TASKS: ______________________________ AREA BOUNDARIES: ___________________

MOUNTING Ht: _____________ FC (high): ________ FC(low):_________

FIXTURE TYPE:___________ LAMP LENGTH: _____ ft LAMP WATTS: _______

MOUNTING: Recessed SM Pend COLOR: CW WW Day LW RE

DEPRECIATION: Max Min Mixed CONDITION: VD Dirty Avg Clean VC
Rusty Peeling Lens Degraded

QTY: ______ BURNOUTS: _________ SWITCHING: Bkr PB/Contactor Sw Tmr

BURN HOURS: Hrs/Wk: ____ Wks/Yr: ___

- - - - - - - - - - - - - - - - - - - - - - - - - - - - - - - - - - - - - - - -

Comments:

- - - - - - - - - - - - - - - - - - - - - - - - - - - - - - - - - - - - - - - -

TASKS: ______________________________ AREA BOUNDARIES: ___________________

MOUNTING Ht: _____________ FC (high): ________ FC(low):_________

FIXTURE TYPE:___________ LAMP LENGTH: _____ ft LAMP WATTS: _______

MOUNTING: Recessed SM Pend COLOR: CW WW Day LW RE

DEPRECIATION: Max Min Mixed CONDITION: VD Dirty Avg Clean VC
Rusty Peeling Lens Degraded

QTY: ______ BURNOUTS: _________ SWITCHING: Bkr PB/Contactor Sw Tmr

BURN HOURS: Hrs/Wk: ____ Wks/Yr: ___

- - - - - - - - - - - - - - - - - - - - - - - - - - - - - - - - - - - - - - - -

Comments:

**Checklist 10.8** Back.

# List of Abbreviations

A — "arbitrary" shape (bulbs)
ANSI — American National Standards Institute
ASHRAE — American Society of Heating, Refrigerating and Air–Conditioning Engineers
BEF — ballast efficacy factor
BF — ballast factor
BR — bulbous reflector
BTU — British thermal unit
CFL — compact fluorescent lamp
C&I — commercial and industrial
CIE — Commission Internationale de l'Éclairage
CLEP — Certified Lighting Efficiency Professional
CLMC — Certified Lighting Management Consultant
CP — candlepower
CRI — color rendering index
CU — coefficient of utilization
DOE — Department of Energy
DSM — demand–side management (programs)
ECM — energy conservation measures
ECOs — Energy Conservation Opportunities
EFF — luminaire efficiency
EIA — (U.S. Dept. of Energy's) Energy Information Administration
ELB — electronic ballast
EPA — Environmental Protection Agency
EPAct — Energy Policy Act of 1992
EPRI — Electric Power Research Institute
ER — elliptical reflector (bulb)
ESCOs — Energy Services Companies
FC — footcandle
FEMP — Federal Energy Management Program
GIGO — "garbage in, garbage out"
GL/DSS — Green Lights/Decision Support System
GWh — gigawatt–hours
HazMat — hazardous material
HID — high–intensity discharge
HO — high output (bulb)
HPS — high–pressure sodium (bulb)
HVAC — heating, ventilating, and air–conditioning
ID — identification
IESNA — Illuminating Society of North America
IRP — integrated resource planning
IRR — internal rate of return
K — Kelvin
kWh — kilowatt–hours
LBF — low ballast factor
LCC — life–cycle costing; life–cycle cost
LCD — liquid crystal display
LDD — luminaire dirt depreciation
LED — light–emitting diode
LER — luminaire efficiency rating
LES — Lighting Evaluation System

LLD — lamp lumen depreciation
LLF — light loss factor
LPW — lumens per watt
mA — milliampere
MBWT — minimum bulb wall temperature
MOU — Memorandum of Understanding (in conjunction with EPA Green Lights program)
MW — megawatts
NALMCO — National Association of Lighting Management Companies
NEC — National Electric Code
NEMA — National Electrical Manufacturers Association
NFPA — National Fire Protection Association
NLPIP — National Lighting Product Information Program
NPV — net present value
OSHA — Occupational Safety and Health Agency
PAR — parabolic aluminized reflector (bulbs)
PC — personal computer; photocell
PCBs — polychlorinated biphenols
PD — power density
PDA — personal digital assistant
PIR — passive infrared
PUC — public utility commission
R — reflector (bulbs)
RCR — room cavity ratio
R&D — research and development
r.m.s. — root–mean–square
ROI — return on investment
RW — reduced wattage
SC — spacing criteria
sf — square feet
SIR — savings–to–investment ratio
SPB — simple payback
STD — "standard" (older coil and core ballast)
T&D — transmission and distribution
TCLP — Toxicity Characteristic Leachate Procedure
THD — total harmonic distortion
TLL — total lamp lumens
TOU — time of use
T.O.W. — twist–of–the–wrist (lamp change)
TSCA — Toxic Substances Control Act
UL — Underwriters Laboratories
UPD — unit power density
UV — ultraviolet
VA — volt–amps
VAC — volts alternating current
VCP — visual comfort probability
VDT — video display terminal
VHO — very high output (bulb)
W/sf — watts per square foot

# Proprietary Names

Calcu–Lite
cal–LIGHT 400
Econo–Watt
ESTIMATOR
Eternalamp
EXCEL
Flex–Lite
GL/DSS
Heath SM–200
HOME CONTRACTOR
idEA$ template
InSite
Lighting Handbook Software
Lighting Logger
Lotus 1–2–3
Measure Master
MicroDataLogger
MyTech Energy Monitor EM–30
Leica DISTO
LightPAD
Merlin
PRO MEASURE
Quikalc
SONIN 45
SONIN 60 PRO
SONIN Combo PRO
Super–Saver
SYLVIA PC
Thrift–Mate
Watt–Miser
Watt Stopper InteliTimer Pro Logger
Windows

# Glossary

**Accent lighting** Directional lighting provided to create a visual focus or to highlight or draw attention to special features such as merchandise displays or architectural features.

**Acrylic** Generic term for a family of quality, light-stabilized plastics used in making luminaire diffusers and lenses.

**Active power** The total input power of a lamp and ballast combination, measured in watts; also, the component of power that provides useful work-producing power. *See* Power.

**Aiming angle** The angle from which the light falls on the surface being lighted. The zero degree aiming angle is perpendicular to the surface being lighted.

**Air-handling** A term applied to a recessed luminaire that supplies or returns comfort air to or from a room. This is usually accomplished through slots along the sides or ends of the luminaire.

**Alternating current (AC)** A periodic current that reverses at regularly occurring intervals of time and has alternately positive and negative values.

**Ambient lighting** Lighting provided throughout an area that is considered uniform general illumination.

**Ambient temperature** The temperature of the air or other medium in the vicinity of an operating lamp, luminaire, or other electrical equipment.

**Ampere (I)** The standard unit of electrical current.

**Angle of incidence** The angle at which a light ray strikes a surface.

**Angle of reflection** The angle at which a light ray leaves a surface.

**ANSI** The American National Standards Institute, Inc. is the coordinator of standards, including lighting standards, on a national level. It is composed of trade organizations, technical societies, professional groups, and consumer organizations, as well as company members

**Apparent power** The root-mean-square summation of three components active power, reactive power, and distortion power.

**Arc lamp** A discharge lamp in which the light is emitted by an arc discharge or by its electrodes.

**Arc tube** A clear quartz or ceramic tube that contains the arc stream. The tube is enclosed within the outer glass envelope of an HID lamp.

**Average rated life** The life value assigned to a particular type of lamp, in hours, at which half of a large group of lamps have failed. The rated life is a median value of life expectancy. Any particular lamp or group of lamps may vary from the published rated life. For fluorescent lamps, the average rated lamp life is affected by the burn cycle. Standard conditions for determining rated life for fluorescent lamps include operating at nominal line voltage and cycling 3 hours on and 20 minutes off. *See also* Effective lamp life.

**Ballast** A current-limiting electrical device used with electric discharge lamps (fluorescent, mercury vapor, metal-halide, high-pressure sodium) to provide the electrical circuit conditions necessary to start and operate the lamp.

**Ballast cycling** Undesirable condition where a magnetic ballast turns lamps on and off (cycles) due to the overheating of the thermal switch inside the ballast. The cause may be incorrect lamps, improper voltage, high ambient temperature at the fixture, or the early stage of ballast failure.

**Ballast efficacy factor (BEF)** The ballast factor divided by the input power of the ballast. The higher the BEF (within the same lamp-ballast type), the more efficient the ballast.

**Ballast factor** The ratio of the lumen output of fluorescent lamp(s) operated on a commercial ballast to the lumen output obtained when operated on the standard (reference) ballast specified for rating lamp lumens.

**Ballast losses** The portion of the power supplied to a ballast that is not transformed into energy used by the lamp but is converted into heat.

**Batwing** One type of light distribution characteristic of fluorescent luminaires, where the maximum candlepower is emitted between 25 and 55 degrees from vertical in a plane perpendicular to the lamp axis.

**Beam lumens** The lumens contained in the beam spread of a floodlight.

**Beam spread** The angle of light distribution from a specific light source on one plane in which the candlepower is 50% of the maximum candlepower.

**BIAX Lamp** *See* compact fluorescent lamp.

**Branch circuit** Circuit power conductors between the last circuit breaker or fuse, in a power panel, and the luminaires. The conductors are run in a raceway, conduit , or flexible wiring system.

**Brightness** As commonly applied, brightness is the intensity of the sensation which results from viewing a surface or space which directs light into the eyes. *See* Luminance.

**British thermal unit (BTU)** A unit of measurement commonly applied as 3.413 BTU being equal to one watt of electrical power.

**Brown out** The electrical power situation when the supply voltage drops below the minimum required by local utility regulations. This low-voltage situation can damage some types of electrical apparatus. For lighting systems, it can extinguish HID lamps and reduce fluorescent light output unless voltage regulation is part of an electronic ballast design. Electric utilities sometimes initiate brown outs to meet demand during peak times.

**Burn cycle** The burning hours per start of a lighting system. Burn cycle affects the average rated lamp life. The average rated life of fluorescent lamps is based on an ANSI test cycle of 3 hours per start. For fluorescent systems that have longer burn cycles, the lamps will have longer than average rated life; for fluorescent systems that have shorter burn cycles, the lamps will have shorter than average rated life.

**Calendar time** *See* effective lamp life.

**Candlepower (CP)** The luminous intensity expressed in candelas. CP is always a property of a source of light and gives information regarding luminous flux at its origin. When describing the distribution of a luminaire, it may be shown in both tabular form and in a polar graph, known as a photometric distribution.

**Capacitor** An electrical device constructed of very thin metallic plates, separated by a dielectric insulator, tightly wrapped in layers, and capable of storing an electrical charge. In AC circuits, when a voltage is applied to the capacitor terminals, capacitive reactance is introduced into the circuit. The capacitance (C) of a capacitor is measured in farads or, more commonly, microfarads.

**Cathode (hot)** In fluorescent lamps, a type of electrode employing a coated coiled coil tungsten wire. The cathode is coated with an emissive material (barium, strontium, and calcium oxide), which when heated gives off electrons by the process of thermionic emission.

**Cathode-disconnect ballast** An electromagnetic ballast that disconnects the electrode-(cathode-) heating circuit. Typically, these ballasts have low ballast factor (BF) that is lowered further when used with reduced-wattage lamps.

**Cavity ratio** The number that indicates the size or proportions of a cavity calculated from its length, width, and height. *See* room cavity ratio (RCR).

**Central control system** An automated system to monitor and control lighting systems from a central computer.

**Certified Ballast Manufacturers (CBM)** A trade association composed of manufacturers who produce lighting ballasts to conform to ANSI specifications C82.1, C82.11, C82.2, and C82.3.

**Chromaticity (of a light source)** The whiteness of a light source expressed in degrees Kelvin (K). Incandescent sources have a dominant chromaticity of 2700K; warm fluorescent sources are rated at 3000K; cool fluorescent sources have a chromaticity of 4100K; and neutral fluorescent sources are rated at 3500K.

**Coefficient of utilization (CU)** The ratio of the lumens from a luminaire received on the work plane to the lumens emitted by the luminaire lamps alone. It is expressed as a percentage or a decimal.

**Coffer** A recessed panel or dome in the ceiling.

**Color rendering index (CRI)** The measure of the degree of color shift that colored objects undergo when illuminated by the light source as compared with the color of those same objects when illuminated by a reference source of comparable color temperature.

**Color temperature** *See* Correlated color temperature.

**Compact fluorescent light (CFL)** The generic name for a family of single-ended fluorescent lamps of folded or bridged tube design with high color rendering (CRI >80) and long life (10,000 hours). CFLs were originally called PL lamps and were originally designed only as preheat lamps, with the starter built into the base. They are often used as alternatives to incandescent lighting. Configurations now include twin tubes, quad tubes, and triple tubes in both preheat and rapid-start models. CFLs are also known as twin-tube or BIAX lamps.

**Constant wattage autotransformer (CWA) ballast** A popular type of HID ballast where the primary and secondary coils are electrically connected. Considered an appropriate balance between cost and performance.

**Constant wattage (CW) ballast** A premium type of HID ballast where the primary and secondary coils are isolated. Considered a high-performance, high-loss ballast, featuring excellent output regulation.

**Contactor** A relay whose contacts are designed to handle lighting loads.

**Contrast** A quality defined by the relationship between the brightness of an object and its immediate background. This contrast between an object and its background enables us to see. *See also* luminance contrast.

**Contrast ratio** The relationship between the luminance of an object and its immediate background, such as the contrast between the letters and the screen background on a video display terminal (VDT).

**Correlated color temperature (CCT)** The absolute temperature, measured in degrees Kelvin, of a blackbody radiator whose chromaticity most nearly resembles that of the light source.

**Counter** A handheld lighting survey tool used to count fixtures.

**Current crest factor (CCF)** *See* Lamp current crest factor.

**Current limiter** *See* Power reducer.

**Cutoff angle (of a luminaire)** The angle, measured up from the nadir, between the vertical axis and the first line of sight at which a reflector, louver, or other shielding device cuts off direct visibility of a lamp. It is the complementary angle of the shielding angle.

**Cycle** A period in which electrical current alternates. A 60-Hz current alternates 120 times per second. *See also* Sine wave, frequency, period, burn cycle, and ballast cycling.

**Cycles per second** *See* Hertz.

**Daylight, daylighting** The light of day; sunlight.

**Daylight harvesting** Lighting control strategy for making effective use of daylighting by using strategically located photosensors to determine the ambient lighting level and adjusting the light output of the lighting system, usually by dimming, to maintain the amount of light set for the space. As daylight levels increase, lamp intensity decreases.

**Demand** The rate at which energy is consumed (kWh/hr), measured in kW.

**Demand control** The lighting control strategy that reduces the peak electrical demand by adjusting the lighting power, usually through dimming or switching.

**Demand interval** The period of time over which the rate of consumption of electrical energy is averaged to obtain peak demand.

**Diffuse** Having the property of scattering or dispersing incident light over a wide range of angles. Materials such as plaster, wood, and brick have diffuse surfaces (similar to having a matte finish).

**Diffuser** Translucent sheet of flat or slightly-bowed, milky-white plastic that scatters light uniformly in all directions below the ceiling plane, reduces the brightness of the source, and shields the light source in a luminaire.

**Dimmer** A control device used to provide variable light output from lamps.

**Direct current (DC)** Electrical current that flows through a circuit in the same direction at all times with constant strength.

**Direct glare** Glare caused by a bright source directly in the field of vision. Often the result of insufficiently shielded light sources.

**Directional lighting** The illumination provided on a work plane, on an object, or on a surface area from a specified direction and location.

**Disability glare** Glare resulting in reduced visual performance and visibility and often accompanied by discomfort.

**Discharge lighting** Light produced from the ionization of gaseous atoms caused by an electric discharge. Common discharge light sources include fluorescent, HID, and low-pressure sodium.

**Discomfort glare** Glare producing discomfort. It does not necessarily interfere with visual performance or visibility.

**Downlight** A type of small direct ceiling luminaire that directs all the light downward; can be recessed, surface-mounted, or suspended.

**Dummy tube** An early energy-management impedance device used with two-lamp ballasts. The dummy tube produces no light and may also reduce the light output of the remaining standard tube. The net result reduces the power requirements of the luminaire by at least 50%.

**Effective** Having an intended or expected effect.

**Effective lamp life** The calendar time between lamp changes, based on lamp burn cycles and average rated lamp life. *See also* average rated lamp life and burn cycle.

**Efficacy** *See* Luminous efficacy.

**Efficiency** *See* Luminaire efficiency and Luminous efficacy.

**Electrode** A coated metal element to facilitate the emission of electrons to form an arc within a fluorescent lamp. *See also* Cathode and Filament.

**Electroluminescent** A light-source technology used in EXIT signs that provides uniform brightness and long life (about 8 years), while consuming little energy (about 1 watt per face)—but suffers from high lumen depreciation.

**Electromagnetic interference (EMI)** High-frequency interference (electrical noise) caused by electronic components or fluorescent lamps that radiate high-frequency waves (while operating) that interfere with normal electronic equipment operation. EMI is measured in microvolts. It can be controlled by filters in the originating equipment. Because EMI can interfere with communication equipment, the Federal Communications Commission (FCC) has established limits for EMI.

**Electronic ballast** *See* high-frequency electronic ballast.

**Electronic dimming ballast** A variable output electronic fluorescent ballast.

**Ellipsoidal reflector (ER) lamp** A reflector lamp in which the filament is located at one of the two focal points of an ellipsoid. All rays of light that strike the ellipsoidal reflector will be reflected through the second focal point. This principle allows light to be focused at a point about 2" in front of the lamp. Sufficient light is directed away from the baffle of a deep downlight to permit a 50% reduction in wattage, yet allow the same light out of the downlight compared to a higher-wattage reflector (R) lamp.

**Energy, electrical** The capacity for doing work, measured in kWh (kilowatt-hours).

**Energy-saving ballast** A type of magnetic ballast designed so that the components operate more efficiently, cooler, and longer than a "standard" magnetic ballast. By U.S. law, standard magnetic ballasts for the most popular types of lamps can no longer be manufactured.

**Energy-saving lamp** *See* Reduced-wattage fluorescent lamp.

**Explosion-proof luminaire** A luminaire that is completely enclosed and is capable of withstanding an explosion of a specific gas or vapor that may occur within it and preventing the ignition of a specific gas or vapor surrounding the enclosure by sparks, flashes, or explosion of the gas or vapor within. It must operate at such an external temperature that a surrounding flammable atmosphere will not be ignited.

**Filament-cutout ballast** *See* Cathode-disconnect ballast.

**Filament (fluorescent)** *See* Electrode and Cathode.

**Filament (incandescent)** *See* incandescent filament lamp.

**Fixture** A luminaire without lamps or means of connecting power.

**Flicker** On lighting systems powered at 60 Hz, magnetically ballasted fluorescent lamps flicker at a frequency of 120 Hz, which some people can consciously detect. Flicker is a known source of eyestrain and fatigue. Fluorescent lamps operated on high-frequency electronic ballasts measure imperceptible flicker rates.

**Floodlight** A luminaire that provides a controlled or concentrated beam of light.

**Fluorescent lamp** A type of electric discharge source in which light is produced predominantly by the fluorescence of phosphors activated by ultraviolet energy from a mercury arc. It consists of a tubular bulb having an electrode (cathode) sealed into each end and containing mercury vapor at low pressure with a small amount of inert gas, usually argon (krypton in reduced-wattage lamps), for starting. The inner walls of the bulb are coated with fluorescent powders. When the proper voltage is applied, a flow of electrons is driven from one electrode and attracted to the other electrode. Collisions between these electrons and mercury atoms in their path cause a state of excitation which results in the emission of radiation, chiefly in the ultraviolet (UV) region at 253 nanometers. The fluorescent coating transforms the UV into visible light.

**Footcandle (FC)** The English unit of illuminance with the foot taken as the unit of length. It is the illuminance on a surface 1 square foot in area on which there is a uniformly distributed flux of 1 lumen.

**Footlambert** The obsolete unit of luminance equivalent to the uniform emittance or reflectance of 1 lumen per square foot from a perfectly diffusing surface and equal to $1/\pi$ candela per square foot. The accepted unit is now candela per square meter. *See* Luminance.

**Glare** The sensation produced by luminances within the visual field that are sufficiently greater than the luminance to which the eyes are adapted to cause annoyance, discomfort, or loss in visual performance or visibility.

**Glossy** Having the reflective properties of a mirror. Example finishes are marble, polished aluminum, and enamel paint.

**Grid (lay-in)** A type of ceiling construction where the supporting members (inverted Tecs) are exposed and the ceiling tiles and luminaires are laid in on the flanges of the Tecs.

**Halogen** Any of a group of five chemically related, nonmetallic elements including astatine, bromine, chlorine, fluorine, and iodine. *See* Tungsten-halogen lamp.

**Harmonic distortion** A harmonic is a sinusoidal component of a periodic wave having a frequency that is a multiple of the fundamental frequency. Harmonic distortion from lighting equipment can interfere with other electronic equipment, as well as the operation of electric power distribution systems. Also, a measure of the difference between the shape of the current wave drawn by a device and the shape of the voltage wave supplied to that device. *See also* Total harmonic distortion.

**Hazardous location** An area where ignitable vapors or dust may cause a fire or explosion by energy emitted from lighting or other electrical equipment.

**Heater-cutout ballast** *See* Cathode-disconnect ballast.

**Heater-cutout lamp** *See* Plus lamp.

**HVAC** Heating, ventilating, and air-conditioning.

**Hertz (Hz)** Unit of frequency; cycles per second (60 cycles per second = 60 Hz).

**Hg** *See* Mercury.

**High-bay lighting** Interior lighting where the roof trusses or ceiling height is greater than 25 feet above the floor.

**High-frequency electronic ballast** A solid-state ballast that provides high-frequency current (in the 20–40 kHz range) to the lamp(s) to increase the system luminous efficacy.

**High-intensity discharge lamp** An electric discharge lamp in which the light-producing arc is stabilized by wall temperature, and the arc tube has a bulb wall loading in excess of 3 watts per square centimeter. HID lamps include groups of lamps known as mercury vapor, metal-halide, and high-pressure sodium.

**High-intensity discharge (HID) lighting** Lighting using mercury vapor, metal-halide, or high-pressure sodium light sources. Although low-pressure sodium lamps are not HID sources, they are often included in the HID category. *See* high-intensity discharge lamp.

**High mast lighting** Roadway or parking lot illumination of a large area by means of a group of luminaires designed to be mounted on the top of a high mast generally 60 feet or higher.

**High output (HO) lamp/ballast** A fluorescent lamp or ballast designed to operate at higher currents (800 mA) and produce more light than smaller wattage lamps and ballasts.

**High power factor (HPF)** The term applied to lighting ballasts and other AC electrical devices that measure 0.9 (90%) or higher. They require the lowest level of current at the specific amount of power needed. This reduces wiring costs by permitting more fixtures on branch circuits. *See* Normal power factor.

**High-pressure sodium lamp** A high-intensity discharge (HID) lamp in which light is produced by radiation from sodium vapor operating at a partial pressure of about 13,300 Pa (100 torr). This category Includes clear and diffuse-coated lamps.

**Hot restart or hot restrike** The method of restriking the arc in an HID light source with a high voltage, after momentary power loss.

**Hybrid ballast** *See* Cathode-disconnect ballast.

**Illuminance (E)** The density of the luminous flux incident on a surface, measured in footcandles or lux; it is the quotient of the luminous flux divided by the area of the surface (lumens per area) when the latter is uniformly illuminated. Illuminance is commonly referred to as light level.

**Illumination** The term commonly used in a qualitative or general sense to designate the act of illuminating or the state of being illuminated. Also an alternative, but deprecated, term for illuminance. Usually the context will indicate which meaning is intended, but occasionally it is desirable to use the expression to indicate that the quantitative meaning is intended.

**Incandescent filament lamp** A lamp in which light is produced by a filament heated to incandescence by an electric current.

**Incident light** The light which falls on a surface.

**Indirect glare** Glare produced from a reflective surface.

**Indirect lighting** Lighting achieved by reflection, usually from wall and ceiling surfaces.

**Inductor** A coil that stores electrical energy when current flows through it and provides inductive reactance in an AC circuit. These coils are also called reactors and produce current that lags the applied voltage. Inductance (L) is measured in henrys or, more commonly, milihenrys.

**Initial light output** Light output in lumens after seasoning the lamp for 100 hours.

**Instant-start** A fluorescent system in which the ballast is designed to deliver an initial high voltage to light the instant-start lamp. The arc current heats the filament by bombardment to provide easy electron emission. No preheating of the filament is required and the starting interval is very short. Instant-start lamps have single-pin bases. Average lamp life of lamps that may be operated either instant-start or rapid-start is shorter by about 20% on instant start ballasts. An exception is the T8 lamp, which has equivalent average rated lamp life rating for instant-start or rapid-start operation for burn hours at 12 hour per start and higher.

**Inverse square law** The law stating that the illuminance on a surface varies directly with the candlepower (CP) of a point source and inversely with the square of the distance (d) between the point source and the surface. ($FC = CP/d^2$)

**Kelvin (K)** A unit of absolute temperature equal to 1/273.16 of the absolute temperature of the triple point of water. This unit is equal to 1 Celsius degree. In lighting it is used as a unit for color temperature (*See* correlated color temperature).

**Kilowatt (kW)** A unit of active power equal to 1000 watts.

**Kilowatt-hour (kWh)** A unit of electrical energy determined by the load (in kW) multiplied by the operating time (in hours).

**Lamp** A generic term for a man-made source of light. By extension, the term is also used to denote sources that radiate in regions of the spectrum adjacent to the visible. In popular use, a lamp is also a lighting unit consisting of a lamp with shade, reflector, enclosing globe, housing, or other accessories. In order to distinguish between the assembled unit and the light source within it, the latter is often called a "bulb" or a "tube" if it is electrically powered.

**Lamp current crest factor (CCF)** A measure of the shape of the lamp current, defined as the peak current divided by the root-mean-square (r.m.s.) current. The CCF is the ratio of the peak current to the effective average (r.m.s.) current. For a pure sine wave, the crest factor is 1 divided by 0.707, or 1.414. CCF is determined by the ballast on which a lamp operates. If the CCF is too high, then lamp life will be shortened and fluorescent tube ends will darken. Many lamp manufacturers will not honor the lamp life warranty if the CCF of the ballast exceeds 1.7.

**Lamp envelope** The shape of either the bare lamp or the capsule surrounding a lamp. Compact fluorescent lamps are commonly available in quad tube, twin tube, triple twin tube, coiled tube, U-shaped tubes, circleline, bullet (capsule), and globe (capsule).

**Lamp life** *See* Average rated lamp life.

**Lamp lumen depreciation (LLD)** The factor commonly used in lighting calculations to account for reduced lumen output over time due to aging of the light source (lamp). LLD is a dimensionless decimal value, representing a percentage of the initial lumen rating.

**Lay-in** *See* Grid.

**Lay-in troffer** *See* Troffer and Grid.

**Lens** A sheet of transparent or translucent glass or plastic (often acrylic) that has some kind of prismatic configuration on the bottom surface to alter the directional characteristics of light passing through it. The prisms may be in the form of pyramids, cones, ridges, or other designs that will refract (bend) transmitted rays of light.

**Light (noun)** 1. Radiant energy that is capable of exciting the retina and producing a visual sensation that is perceived by the normal, unaided human eye. The visible portion of the electromagnetic spectrum that extends from about 380 nanometers (violet) to 770 nanometers (red) (3800–7700 angstroms). 2. The sensation of perceiving light; brightness. 3. A source of light, especially a lamp or a luminaire. 4. The illumination derived from a source of light, such as the light of the moon. 5. A particular quantity or quality of such illumination. 6. Daylight; dawn; daybreak.

**Light (verb)** 1. To cause to give out light; make luminous; light a lamp. 2. To provide, cover, or fill with light; illuminate. 3. To emit light; to be lighted. 4. To signal, direct, or guide with or as if with illumination.

**Light-emitting diode (LED)** A p-n junction solid-state diode whose radiated output is a function of its physical construction, material used, and exciting current. The output may be in the infrared or in the visible region. Some low-wattage EXIT lights employ arrays of LEDs to form the letters E-X-I-T. Other EXIT lights and EXIT light retrofits employ rows of LEDs to illuminate the interior of the unit, and still others use LEDs in glass bulbs similar in shape to the incandescent bulbs they replace.

**Lighting control** Electrical devices or systems used to control the time that lighting systems are operated or to control their light output.

**Lighting distribution** Luminaires are classified according to the manner in which they control or distribute the luminous flux.

**Lighting power budget** As applied to a space—the amount of power that may be used by the lighting system in a given space. As applied to a building—the total amount of power that may be used by the lighting systems within the building. *See* Unit power density.

**Light level** *See* Illuminance.

**Light loss factor (LLF)** The factor used in lighting calculations to obtain maintained illuminance (footcandles). It is the composite of eight separate factors (multiplied together) that account for both recoverable and nonrecoverable light depreciation factors. The recoverable factors are lamp burnout factor, lamp lumen depreciation (LLD), luminaire dirt depreciation (LDD), and room surface dirt depreciation (RSDD). The nonrecoverable factors are luminaire ambient temperature, voltage to luminaire, ballast factor, and luminaire surface depreciation.

**Light trespass** A situation that occurs when, due to lack of adequate beam control, light from an outdoor source illuminates areas where the illumination is not wanted.

**Logger** A lighting survey instrument that measures hours-of-use.

**Louver** A series of baffles, arranged in a geometric pattern, used to control light distribution from a luminaire and to shield the source from view at certain angles or to absorb unwanted light. Types can range from small-cell plastic louvers to large-cell anodized aluminum louvers used in parabolic fluorescent luminaires.

**Low-bay lighting** Interior lighting where the roof trusses or ceiling height is approximately 25 feet or less above the floor.

**Low-pressure sodium lamp** A discharge lamp in which light is produced by radiation from sodium vapor operating at a partial pressure of 0.1 to 1.5 Pa. Considered a monochromatic light source.

**Low-voltage lamp** A lamp, typically small tungsten-halogen, that provides both intensity and good color rendition. Low-voltage lamps operate at 12 volts and require a transformer or electronic power supply. Popular lamps include MR11, MR16, and PAR36.

**Low-voltage relay** The relay part of a 24-volt low-voltage switching system; usually a mechanically latching type relay that is operated by momentarily operating a low-voltage switch or other low-voltage system device that permits remote control of lighting.

**Low-voltage switch** The switch part of a 24-volt low-voltage switching system; usually a momentary, rocker-type switch that momentarily pulses a low-voltage relay on and off.

**Lumen** The SI unit of luminous flux. Lumens represent the time rate of transferring luminous energy; quantity of light. One lumen is the light flux falling on a surface of 1 square foot, every part of which is 1 foot from a point source having a luminous intensity of 1 candela in all directions.

**Lumen depreciation compensation** The lighting control strategy that sets the initial light level below maximum lamp output and then gradually increases the light output to compensate for lamp lumen depreciation.

**Lumen maintenance** *See* Lamp lumen depreciation.

**Luminaire** A complete lighting unit consisting of a lamp or lamps, together with the parts designed to distribute the light, position and protect the lamps, and connect the lamps to the power supply.

**Luminaire dirt depreciation (LDD)** The light loss factor that describes how light is lost from the initial illuminance provided by clean, new luminaires to the reduced illuminance that they will provide due to dirt collection on the luminaire surfaces at the time at which it is expected that cleaning will be done.

**Luminaire efficiency** The ratio of lumens emitted by a luminaire to that emitted by the lamp(s) alone expressed by a percentage. If two luminaires use the same lamps, more light will be emitted from the luminaire with the highest efficiency.

**Luminaire surface depreciation** The light loss factor that describes how light is lost as the reflecting surfaces of a luminaire degrade over time.

**Luminance** The quality of being luminous. Light emitted or reflected from a source or object in a particular direction. Luminance produces the sensation of brightness and is measured in candelas per square foot (or per square meter) of source or object area in the direction of viewing.

**Luminance contrast** The relationship between the luminances of an object and its immediate background.

**Luminous efficacy** A measure used to compare light output to energy input; the quotient of the total luminous flux emitted divided by the total lamp power input. Efficacy is measured in lumens per watt. For example, if a 100-watt light source produces 1750 lumens, then the efficacy of the light source is 17.5 lumens per watt.

**Luminous flux** *See* Lumen.

**Luminous intensity** The luminous flux per unit solid angle produced by a light source in a given direction.

**Lux** The SI unit of illuminance. One lux is one lumen per square meter.

**Maintenance factor** *See* Light loss factor.

**Matte** A dull, often rough finish.

**Maximum ambient temperature** The maximum air temperature for which the product is warranted to achieve rated life.

**Maximum overall length (MOL)** The length of a lamp (or the capsule surrounding it), measured from the top of the product to the bottom of the screwbase, in inches.

**Mean light output** The light output of a light source in lumens at 40% of rated lamp life. Also called design light output.

**Mercury (Hg)** A silvery-white poisonous metallic element that is liquid at room temperature; used in thermometers, vapor lamps (fluorescent and HID), and batteries. Atomic number 80; atomic weight 200.59.

**Mercury (vapor) lamp** A high-intensity discharge (HID) lamp in which the major portion of the light is produced by radiation from mercury operating at a partial pressure in excess of 100,000 Pa (1 atmosphere). Mercury vapor lamps emit a blue-green cast of light and have low CRI. Includes clear, phosphor-coated, and self-ballasted lamps.

**Metal-halide lamp** A high-intensity discharge (HID) lamp in which the major portion of the light is produced by radiation of metal halides and their products of dissociation, possibly in combination with metallic vapors such as mercury. Includes clear and phosphor-coated lamps.

**Minimum bulb wall temperature (MBWT)** The coldest point on the lamp bulb wall of a fluorescent lamp. Light output is a function of mercury vapor pressure, which is directly related to MBWT. The optimal MBWT typically occurs at 25°C which is the temperature used in lamp standard test conditions.

**Mounting height** The vertical distance between the luminaire and the surface to be lighted. For exterior, pole-mounted luminaires, it includes the pole length and the base (above grade) to which the pole is mounted.

**MR11 lamp** A low-voltage quartz (mirrored reflector) lamp, 1.375 inches in diameter. The lamp and reflector are one unit, which directs a precise beam of light.

**MR16 lamp** A low-voltage quartz (mirrored reflector) lamp, 2 inches in diameter. The lamp and reflector are one unit, which directs a precise beam of light.

**Nonuniform lighting** A lighting system containing luminaires that are located with respect to individual tasks so more light reaches tasks rather than surrounding areas.

**Normal power factor (NPF)** The term used for low power factor ballasts and other AC electrical devices that draw about twice as much current as high power factor ballasts. This limits the number of fixtures by approximately one-half that can be wired to each lighting circuit. Many low-wattage fluorescent systems use normal power factor ballasts, as well as many compact fluorescent units that use simple choke ballasts. *See* High power factor.

**Occupancy recognition** The lighting control strategy applied to intermittently occupied areas or rooms, to turn lights on when people are present and off automatically after a certain amount of time when they are no longer present.

**Occupancy sensor** The lighting control that turns lights off after a space becomes unoccupied. Used to implement occupancy recognition control strategies.

**Optical System** The lamp cavity or environment (including diffusing media) of a luminaire designed to control the light output.

**Optics** The term referring to the optical components of a luminaire, such as reflectors, refractors, lenses, or louvers, or to the light-emitting or light-controlling performance of the luminaire.

**Outdoor lighting** Lighting used to light areas outside a building, including roadways, walkways, parking lots, landscapes, and building facades.

**PAR 16** A PAR lamp that is 16 one-eighths of an inch (2") in diameter. Other PAR lamp sizes are PAR 20 (2.5"), PAR 30 (3.75"), PAR 36 (4.5"), PAR 38 (4.75"), PAR 46 (5.75"), PAR 56 (7"), and PAR 64 (8").

**Parabolic aluminized reflector (PAR) lamp** A focused lamp with accurate beam control designed into the lamp. It consists of a hard glass envelope with an inside coated reflecting surface, a precisely placed filament at the focal point of the parabolic reflector, and a lens to control beam spread.

**Parabolic curve (paraboloid)** A geometric shape (or a section) used in the optics of luminaires to control light. The paraboloid is a three-dimensional parabola and has the property of directing light parallel to its geometric axis when a light source is placed at its focal point.

**Parabolic louver** Parabolic luminaires use large-cell louvers formed in a parabolic shape. The resultant light distribution reduces glare, controls light output, and has high aesthetic appeal.

**Parabolic luminaire** A popular type of fluorescent luminaire that has a louver composed of aluminum baffles that are curved in a parabolic shape. The resultant light distribution produced by this shape provides reduced glare, better light control, and is considered to have greater aesthetic appeal.

**Parabolic reflector** A light reflector shaped as a paraboloid to concentrate a light beam parallel to the axis of the reflector.

**Paracube** A metallic coated plastic louver constructed of small squares (1/2-inch squares are common); often used to replace the lens in an installed troffer to enhance its appearance. Small cell louvers have high visual comfort probability (VCP), but sacrifice luminaire efficiency and spacing criteria (SC) to achieve it.

**Passive infrared (PIR)** One of the principal technologies used in occupancy sensors. Passive infrared sensors react only to energy sources as they move between the fan pattern detection zones created by the lens of the sensor.

**Peak demand** The maximum rate at which energy is consumed (kWh/hr), measured in kilowatts (kW).

**Pendant-mounted luminaire** A luminaire supported from the ceiling structure by stems or pendants which position the luminaire below the finished ceiling.

**Photocell** An electrical device that switches lighting loads in response to changes in ambient light. The sensor is usually a cadmium-sulfide cell that operates a set of contacts to control the lighting load. Used for dusk-to-dawn switching of exterior lighting systems.

**Photometric brightness** A term formerly used for luminance.

**Photometric distribution** A polar graph representing the variation in candlepower (luminous intensity) of a luminaire in a given reference plane.

**Photometric report** A report generated from laboratory testing, usually by a luminaire manufacturer or an independent testing laboratory. The report contains a table of printed data describing the light distribution, the polar graph photometric distribution, luminaire efficiency, zonal lumen outputs, and a coefficient of utilization table for the luminaire under test.

**Photosensor** An electrical device that produces an output signal that is an analog of the light measured on the silicon cell. Used for daylight-harvesting and lumen depreciation control strategies.

**Plenum** The space between the structural ceiling slab and the finished ceiling. The space may contain air ducts, electrical wiring, sprinkler piping, etc. It is the space that conceals the housing part of a recessed luminaire.

**Plus lamp** A 4-foot reduced-wattage fluorescent lamp with a thermal-activated switch that cuts the power to each of the lamp's cathodes to reduce the lamp watts to 32 watts. The thermal switch provides a restrike time of about 1 minute. These lamps are not compatible with electronic ballasts or power reducers and exhibit poor performance with occupancy sensors. Brand names are Super-Saver Plus and Watt-Miser Plus.

**Point source** A light source with dimensions less than about one-fifth the distance between the source and the surface to be illuminated.

**Polychlorinated biphenyl (PCB)** A dielectric material used in ballast capacitors that is classified as a hazardous waste material (HazMat) that must be disposed of by an approved method.

**Polystyrene** A family of plastics used in the manufacture of luminaire diffusers and lenses. Tends to yellow or appear brown due to the effect of ultraviolet radiation from fluorescent and HID lamps.

**Position factor** The ANSI standard testing environment for rating compact fluorescent lamp light output is with the compact fluorescent lamp operating in a vertical base-up position. For other orientations, light output will differ. Position factor is defined as the light output produced by the lamp in a certain orientation divided by the light output produced by the lamp in the base-up position.

**Potting** The material used as a filler in magnetic and some electronic ballasts to dissipate heat and reduce noise. In magnetic ballasts the filler material is usually a mixture of tar and sand.

**Power** The time rate at which electric energy is used, commonly measured in watts or kilowatts (kW).

**Power density** *See* Unit power density.

**Power factor (PF)** The ratio of active power, in watts, to the apparent power, in r.m.s. volt-amperes. It is a measure of the efficiency with which an AC electrical device converts input current and voltage into useful electrical power. A device such as a ballast that operates at 120 volts line voltage, draws 1 amp, and consumes 60 watts has a power factor of 0.5 or 50%. Some electric utilities charge customers for low PF systems.

**Power line carrier** The lighting control medium that uses transmitters to couple control signals to existing power lines that are used to carry the signals to receivers, also connected to the same power lines. Used in control projects. *Note* Low harmonics electronic ballasts interfere with older PLC equipment operation and are incompatible.

**Power quality** Refers to the effects that lighting systems may have on facility and utility electrical distribution systems.

**Power reducer** An electrical apparatus that is an impedance element placed in series between a magnetic ballast and fluorescent lamps. Reduces both power and lumen output.

**Preheat** The original fluorescent system that uses a separate starter or a manual switch to preheat the fluorescent lamp filaments. After the filaments are heated, the starter opens and the ballast then provides a suitable voltage to light the lamp and limit the lamp current to the proper value. Several seconds are required to complete the starting operation. Early low-wattage compact fluorescent lamps are preheat and have a starter built into the base.

**Prismatic lens** A sheet of glass or plastic consisting of prisms that refract transmitted light.

**Quad-tube lamp** A compact fluorescent lamp with a double twin-tube configuration.

**Quartz lamp** *See* Tungsten-halogen lamp.

**Raceway** A tube or channel that encloses and protects electric wires.

**Radio frequency interference (RFI)** Interference ("noise") in the radio frequency band caused by other high-frequency equipment or devices in the immediate area. Fluorescent and HID lighting sources generate RFI.

**Rapid-start** The fluorescent system in which the ballast provides the proper energy level to heat the lamp filaments continuously through small low-voltage filament windings. Rapid-start operation reduces the open circuit voltage requirements for starting and speeds up the starting interval. Usually a grounded reflector close to the lamp, together with a grounded line circuit, is required to provide a capacitive effect between lamps and fixture body and to assist ionization in the starting process.

**Reactor** *See* Inductor.

**Recessed** A term describing a luminaire mounted in a ceiling opening so that the housing of the luminaire is hidden from view. The luminaire's lens/door frame may slightly protrude, may be flush, or may be slightly regressed relative to the ceiling surface.

**Reduced-wattage fluorescent lamp** A lower-wattage retrofit lamp, producing fewer lumens, that also reduces light output by reducing ballast factor on magnetic ballasts. Must be operated above 60°F to prevent starting and operating problems. Not compatible with all electronic ballasts because the tin oxide coating under the phosphor coating tends to short out the high-frequency current.

**Reflectance** The ratio of the light reflected from a surface to the light falling on it.

**Reflected glare** Glare from specular reflections of high brightness off polished or glossy surfaces.

**Reflection** A general term for the process by which the incident flux leaves a surface or medium from the incident side, without change in frequency. Reflection is usually a combination of specular and diffuse reflection.

**Reflector** A device used in some luminaire designs to redirect light from a light source by using the principle of reflection. Also A retrofit product to upgrade the performance of fluorescent luminaires by increasing the interior reflectance.

**Reflector (R) lamp** A directional incandescent lamp with a frosted coating on the face of the lamp to control beam spread. Many older R lamps are no longer manufactured or imported, due to provisions of the Energy Policy Act of 1992 (EPAct).

**Refraction** The phenomenon that takes place when light rays change direction as they pass obliquely from one medium (air) to another (glass or plastic).

**Refractor** A device used in some luminaire designs to redirect light by using the principle of refraction.

**Regulation** Describes the degree to which a ballast holds the output (watts and light) constant during fluctuations in the input voltage to the ballast. Normally specified as +/− percent change in output compared to +/− percent change in input voltage.

**Relay** An electrical device having an electromagnetic coil that, when energized, closes its contacts to complete an electrical circuit. *See also* Low-voltage relay.

**Retrofit** Upgrade or improvement in the performance of lighting accomplished by improving the electrical, optical, or thermal efficiency of lighting equipment and/or improving the utilization of light by improving rooms and buildings in which the light is provided.

**Rheostat** An obsolete lighting control using a variable resistor to dim incandescent lighting systems. It is no longer used, because power is wasted in the resistive element of the rheostat. Solid-state dimmers using electronic switching have replaced rheostats, and do not waste power.

**Roadway lighting** The fixed lighting of public ways that supplements vehicular headlight illumination to allow motorists and pedestrians to see in a nighttime environment.

**Room surface dirt depreciation (RSDD)** The percentage loss of task illuminance due to dirt accumulated on room surfaces over time.

**Root-Mean-Square (r.m.s.)** The square root of the mean of the sum of the squares of instantaneous amplitudes of a waveform. The root-mean-square current (or voltage) in an alternating current is the effective average value of the current (or voltage) that provides the equivalent heating effect as a DC power source.

**Scheduling** A time-based lighting control strategy to activate, extinguish, or adjust lighting according to a predetermined schedule.

**Semi-specular** Term describing the light reflection characteristics of a material. Some of the light is reflected directionally, with some amount of scatter (diffusion).

**Shielding angle (of a luminaire)** The angle measured between a horizontal line drawn through the light center and the line of sight at which the bare source first becomes visible. The shielding angle is the complementary angle of the cutoff angle. Higher shielding angles reduce direct glare.

**Shielding media** Material located in front of the lamps to redirect the light and/or reduce luminaire brightness. Lens, diffusers, and baffles are all examples of shielding media.

**Spacing criteria (SC)** The information regarding how far apart luminaires may be spaced (center-to-center) to maintain uniformity of illumination on the work plane. To use the spacing criteria, multiply the net mounting height (luminaire to work plane) by the spacing criteria value. The resulting number represents the maximum (center-to-center) distance that the luminaires may be spaced to maintain uniformity.

**Spacing-to-mounting-height ratio (S/MH)** The ratio of the distance between the center point of luminaires to their mounting height above the work plane.

**Specular** The term used to describe the finish of materials used in some louvers and reflectors used in luminaires. These materials are usually mirrored or polished surfaces that reflect light. The angle of reflection is equal to the angle of incidence. *See also* Semi-specular.

**Spill light** Light distributed by an outdoor reflector lamp or luminaire that is outside the beam spread.

**Starting method** The method used in a ballast design to start the lamp; instant start, preheat, rapid-start, or modified rapid-start.

**Static troffer** A troffer with no air-handling provisions.

**Stroboscopic effect (strobe effect)** The condition where rotating machinery or other rapidly moving objects appear stationary due to the low-frequency alternating current (AC) supplied to light sources.

**Styrene** *See* Polystyrene.

**Sunshine** The light or the direct rays from the sun.

**Surface-mounted luminaire** Any luminaire mounted directly on the ceiling.

**Switching** The lighting control strategy associated with turning lighting on and off. The common wall switch is the simplest form.

**T8, T10, T12 lamps** Industry standard tubular (T) fluorescent lamps that respectively measure 8 one-eighths of an inch (1.0"), 10 one-eights of an inch (1.25"), and 12 one-eighths of an inch (1.5") in diameter.

**Tandem wiring** A wiring option in which a ballast is shared by two or more luminaires. The benefits of tandem wiring include lower labor, materials, and energy costs. Sometimes tandem wiring is referred to as "master-slave" or "master-satellite" wiring.

**Thermal factor** The thermal environment surrounding fluorescent lamps affects the mercury vapor pressure in the lamp, and thus its light output, by influencing the minimum bulb wall temperature (MBWT). Each lamp has an optimal vapor pressure, and thus MBWT, that provides maximum light output. This optimal MBWT typically occurs at 25°C ambient temperature, which is the temperature used in the standard test conditions. Operating fluorescent lamps above or below the optimum temperature will reduce light output. (A study of compact fluorescent lamps suggests that elevated temperatures will cause light losses between 15 and 20% due to thermal effects.) (*See* also MOWT).

**Third harmonic distortion** The component of a periodic wave having a frequency three times the fundamental. In a 60-Hz wave, the third harmonic is 180 Hz.

**Time control** Also known as time clocks or time switches, time controls activate and deactivate their lighting loads at user-determined times. They are available as electromechanical or electronic devices.

**Torchere** An indirect floor lamp that sends all or nearly all of its light upward.

**Total harmonic distortion (THD)** The ratio of an electric wave's harmonic content to its fundamental component, expressd as a percentage. It is a measure of the degree to which a waveform is distorted by harmonic content.

**Translucent** Transmitting light but causing enough diffusion to prevent the perception of distinct images.

**Transparent** Capable of transmitting light so that objects or images can be seen as if there were no intervening material.

**Trigger-start** A method of starting some low-wattage (15- and 20-watt) preheat fluorescent lamps from a ballast called a trigger-start ballast that also contains the starter.

**Troffer** Recessed—A luminaire that is mounted within the ceiling in such a manner that its bottom surfaces are in the same plane as the ceiling. Surface-mounted—A luminaire that is attached directly on the ceiling and projects below the ceiling by the depth of the luminaire. The term is derived from "trough" and "coffer."

**Tungsten-halogen lamp** A gas-filled tungsten incandescent lamp containing a certain proportion of halogens in an inert gas whose pressure exceeds 3 atmospheres. Examples The tungsten-iodine lamp (UK) and the quartz-iodine lamp (USA).

**Twin-tube** *See* Compact fluorescent lamp.

**Ultrasonic** Above the normal range of human hearing. One of the principal technologies used in occupancy sensors. Ultrasonic sensors are volumetric detectors that employ the Doppler principle, detecting movement by measuring the time between the transmission of signals and their return.

**Ultraviolet (UV) radiation** For practical purposes, any radiant energy within the wavelength range 10 to 380 nanometers. It is invisible radiation that is shorter in wavelength than visible violet light (literally beyond the violet light). UV is emitted when the outer jacket of some HID lamps is broken.

**Underwriters Laboratories (UL)** An independent organization whose responsibilities include rigorous testing of electrical products. Products that pass UL tests can be labeled "UL listed" and can be advertised as such. UL tests only for product safety, including electrical safety.

**Uniform lighting** A lighting system consisting of luminaires located in a uniform grid pattern so that most areas of the space receive nearly the same illumination.

**Unit power density (UPD)** The lighting power allowance for a building or space, measured watts per square foot.

**Valence** A longitudinal shielding member mounted across the top of a window or along a wall. A valence is usually parallel to the wall, to conceal light sources, giving both upward and downward distributions.

**Vandal-resistant luminaires** Luminaires with rugged housings, break-resistant type shielding, and tamperproof screws.

**Vapor-tight luminaire** A luminaire designed and approved for installation in damp or wet locations. Also described as "enclosed and gasketed."

**Veiling reflections** Regular reflections superimposed upon diffuse reflections from an object that partially or totally obscure the task details to be seen (because the contrast is reduced). Sometimes called reflected glare.

**Very high output (VHO) lamp** A fluorescent lamp that operates at a "very high" current (1500 mA), producing more light output than a "high output" lamp (800 mA) or a standard output lamp (430 mA).

**Video display terminal (VDT)** A computer or cathode ray tube screen on which is displayed text or graphical information to be viewed by a person.

**Visibility** The quality or state of being perceivable by the eye. In outdoor applications it is defined in terms of the distance at which an object can be just perceived by the eye. In indoor applications it is defined in terms of the contrast or size of a standard test object, observed under standardized viewing conditions, having the same threshold as the given object.

**Visual comfort probability (VCP)** The subjective rating of a lighting system expressed as percentage of occupants who, when viewing from a specified location and in a specified direction, will be expected to find it acceptable in terms of discomfort glare. VCP takes into account luminaire luminances at different angles of view, luminaire size, room size, luminaire mounting height, illuminance, and room surface reflectances. VCP tables are sometimes included in photometric reports.

**Visual perception** The interpretation of impressions transmitted from the retina to the brain in terms of information about a physical world displayed before the eye. Involves any one or more of the following recognition of the presence of something; identifying it; locating it in space; noting its relation to other things; identifying its movement, color, brightness, or form.

**Visual performance** The quantitative assessment of the performance of a task taking into consideration speed and accuracy.

**Visual task** The term that conventionally designates those details and objects that must be seen for the performance of a given activity, and includes the immediate background of the details or objects.

**Volt** The standard unit of measurement for electrical potential.

**Voltage** The difference in electrical potential between two points of an electrical circuit.

**Voltage factor** The fractional (or percentage) loss of task illuminance due to out-of-tolerance voltage at the luminaire.

**Watt** A unit of power.

**Work plane** The plane (or level) at which work is usually performed and on which the illuminance is specified and measured. Unless otherwise specified, this is assumed to be a horizontal plane 30 inches (2.5 feet) above the floor. In corridors (hallways), the work plane is the floor.

**Wraparound luminaire** A surface- or pendant-mounted luminaire with a one-piece plastic lens that encloses the lamp compartment on both sides and across the bottom, literally wrapping the lamps.

**Zenith** The direction directly above a luminaire (180 photometric degrees).

**Zonal-cavity method** The zonal-cavity inter-reflectance method is a procedure for calculating coefficients of utilization (CU), wall luminance coefficients, and ceiling cavity luminance coefficients, taking into consideration the luminaire intensity distribution, room size and shape (cavity ratio), and room reflectances. It is based upon flux transfer theory.

# Index

## M



## N



## O

## W